European Agency for Safety an

Gender issues in
safety and health at
work
A review

European Agency
for Safety and Health
at Work

R E S E A R C H

Gender issues in safety and health at work — A review

Contributors:

Topic Centre on Research — Work and Health members:

Kaisa Kauppinen and Riitta Kumpulainen, FIOH, Finland
Irene Houtman, TNO Work and Employment, Netherlands

European Agency for Safety and Health at Work: Sarah Copsey

In cooperation with:
Topic Centre on Research — Work and Health members:
Anneke Goudswaard, TNO Work and Employment, Netherlands
Maria Castriotta, ISPESL, Italy
Alan Woodside, OSHII, Ireland
Birgit Aust, AMI, Denmark
Veerle Hermans, Prevent, Belgium
Dolores Solé, INSHT, Spain
Karl Kuhn BAuA, Germany
Ellen Zwink, BAuA, Germany

A great deal of additional information on the European Union is available on the Internet. It can be accessed through the Europa server (http://europa.eu.int).

Cataloguing data can be found at the end of this publication.

Luxembourg: Office for Official Publications of the European Communities, 2003

ISBN 92-9191-045-7

© European Agency for Safety and Health at Work, 2003
Reproduction is authorised provided the source is acknowledged.

Printed in Belgium

FOREWORD

Achieving gender equality in all aspects of employment is now a key European priority. It is a matter of rights, but it is also a matter of sound economic policy — especially considering the human and economic costs of injuries and ill health caused or made worse by work.

The European Commission has adopted a comprehensive policy ([1]) to tackle gender inequalities, based on integrating or mainstreaming gender into all its activities. On this basis, the Community strategy on health and safety at work for 2002–06 ([2]) stresses that measures should be taken in order that gender equality is mainstreamed throughout the strategy. This report, by investigating issues such as the extent to which gender is taken into account in the occupational safety and health field, and gender differences in risk exposure and prevention, should make an important contribution to taking gender policy forward in this area.

It offers an insight into the risks women face at work, how these arise, and how their situation compares to their male colleagues. It provides an analysis of existing research information, practical information on improving risk prevention by taking account of gender issues and positive proposals for the policy level to consider.

The report highlights the dual importance of considering gender in risk prevention and including occupational safety and health in gender equality employment activities. Cooperation between these two policy areas is crucial, from the European level, down to the workplace, to promote improved workplace risk prevention for both women and men.

The Commission believes that this report will be an invaluable catalyst for advancing discussions with all key players (European institutions, national authorities, social partners, and experts) and for developing concrete measures to ensure that the gender dimension is effectively integrated into health and safety policies.

Anna Diamantopoulou

European Commissioner responsible for employment and social affairs

---

([1]) Commission communication COM(2000) 335.

([2]) Commission communication COM(2002) 118.

PREFACE

The world of work is changing. There has been a shift away from rural work and also away from heavy industry and manufacturing. New jobs are being created in the service sector and information technological change has created new types of work and forms of working. Globalisation has increased competition and pressures to rationalise and be more flexible. One important change has been the entry of more women into the labour market.

Prevention of work-related deaths, injuries and ill health is important, because of the high costs of failure to do so to enterprises, to society and to the individuals concerned. Continued efforts to improve prevention for both male and female workers are therefore needed. However, the changes mentioned above and employment conditions and circumstances of life outside work can have different implications for working women and men, including with regard to their work-related health. It is EU policy to promote equality of men and women in all areas that it covers, including occupational safety and health. In the light of this and the increasing participation rate of women in the EU labour market, and the EU's aim to increase this participation still further, 'mainstreaming' the gender dimension into occupational safety and health was included as one of the key objectives in the European Community strategy on safety and health at work 2002–06.

To contribute to this European objective, the Administrative Board of the European Agency for Safety and Health at Work decided to include a report to review gender issues in occupational safety and health in the Agency's 2002 work programme.

The aim of this report is not only to give a clear overview of gender differences in safety and health at work and how they arise, but also to provide information about what this means for prevention and how a gender-sensitive approach can be taken in occupational safety and health.

In this way, we hope that the report can give strong guidance on what is needed to mainstream gender into all areas of occupational safety and health in practice and will serve as an important input to the realisation of the Community strategy, which includes gender mainstreaming as an objective.

The Agency would like to thank Kaisa Kauppinen and Riitta Kumpulainen from FIOH, Irene Houtman and Anneke Goudswaard from TNO, Maria Castriotta from ISPESL, Alan Woodside from OSHII, Birgit Aust from AMI, Veerle Hermans from Prevent, Dolores Solé from INSHT, and Karl Kuhn and Ellen Zwink from BAuA for their contributions to the drafting of this report. The Agency would also like to thank its focal points and other network group members and all other contributors for their valuable comments and suggestions with respect to the report.

European Agency for Safety and Health at Work

November 2003

European Agency for Safety and Health at Work

Contents

EXECUTIVE SUMMARY	9
1. INTRODUCTION	19
Why study health and safety at work from a gender perspective?	19
Key questions to be addressed	20
What to include — the conceptual model	20
The methodology	22
Concepts and terminology	22
A word on statistics	22
2. GENDER DIFFERENCES IN WORKING LIFE IN EUROPE	25
Introduction	25
Women's participation in the labour market	26
Working time	26
Employment status	28
Gender differences in unpaid household and caring tasks	29
The impact of changes in the world of work	30
Gender pay gap	32
Gender segregation in occupations	32
Conclusion	34
3. GENDER DIFFERENCES IN OCCUPATIONAL EXPOSURES AND OUTCOMES	35
Introduction	35
Accidents	35
Musculoskeletal disorders	40
Work-related stress	46
Injuries and stress arising from work-related violence from members of the public	58
The effects of sexual harassment from colleagues	62
The effects of workplace bullying	63
Work and domestic violence	65
Infectious diseases	66
Asthma, other respiratory disorders, and sick-building syndrome	67
Skin diseases	69
Work-related cancer	70
Hearing disorders	74
Vibration-related diseases	77
Working temperatures	78
Reproductive health	79
The effects of working hours and inflexibility, and responsibility for household duties	85
Occupational voice loss — an emerging risk	89
Appearances	90
The health and safety of women in agriculture	91
Health and safety in the fisheries sector	92
The health and safety of domestic workers and homeworkers	93
Occupational safety and health of women working in 'non-traditional' areas	96

7

	Working in SMEs	97
	Occupational safety and health of commercial sex workers	99
	The work-related health of older women	100
	Migrant women workers	101
	Occupational hazards and gender worldwide	101
	Different jobs, different exposures — implications of gender segregation	101
4.	ABSENTEEISM, DISABILITY, COMPENSATION AND REHABILITATION	105
	Sickness absence	105
	Compensation for occupational injury and ill health	107
	Rehabilitation into work	107
5.	GENDER ISSUES IN RESEARCH, LEGISLATION, RESOURCES, SERVICES AND PRACTICE	109
	Introduction	109
	Information gathering for research and statistical monitoring	109
	Information and support for workplace activities	112
	Access to and functioning of occupational health services	115
	Consultation and participation of women in workplace occupational safety and health matters	116
6.	GENDER MAINSTREAMING	123
	Occupational safety and health legislation	123
	EU strategy on occupational safety and health and gender	126
	Mainstreaming, gender and occupational safety and health	127
	Mainstreaming occupational safety and health into equality initiatives	128
	Gender equality in and outside the workplace	130
7.	DISCUSSION AND CONCLUSIONS	131
8.	RECOMMENDATIONS	139
9.	GLOSSARY	143
10.	REFERENCES	145
ANNEXES		177
Annex 1:	Acknowledgements	178
Annex 2:	Project organisation, participants and experts	179
Annex 3:	Extracts relevant to gender from the Community strategy on health and safety at work 2002–06	181
Annex 4:	Extracts relevant to gender from the European Parliament resolution on the Community strategy on health and safety at work 2002–06	183
Annex 5:	Statistical data related to gender and employment in the EU	185
Annex 6:	Data from the third European survey on working conditions	191
Annex 7:	Data from Eurostat on the health and safety of men and women at work	197
Annex 8:	Occupational hazards and gender worldwide	203
Annex 9:	Summary of the relationship between gender, occupational status and working conditions: findings of a gender analysis of the third European survey on working conditions	211
Annex 10:	Member State differences in social security sickness absence and disability regulations	214
Annex 11:	Results from the TUTB survey on the gender dimension in health and safety, and examples of trade union initiatives	216
Annex 12:	Gender equality legislation and policy in the EU	219
Annex 13:	Criteria for assessing workplace equality	220
Annex 14:	Quality of women's work: strategies for change identified by the European Foundation	221

EXECUTIVE SUMMARY

The need to examine gender in occupational safety and health

Women's share of the workforce grew in most EU countries between 1990 and 2000, and today women make up 42 % of the EU workforce, although there are differences between the Member States and particularly between the north and south. Along with the increasing participation rates, new types of work practices and forms of work bear witness to an unmistakable period of change. Such change, as well as different labour, social and health policies, can all impact differently on the working lives of women and men. Cultural differences must also be added to the equation.

Because of strong occupational gender segregation in the EU labour market, which remains high despite changes in the world of work, women and men are exposed to different workplace environments and different types of demands and stressors even when they are employed by the same sector and ply the same trade. There is strong segregation between sectors, between jobs in the same sector, and there can be segregation of tasks even when women and men have the same job title in the same workplace. There is also strong vertical segregation within workplaces, with men more likely to be employed in more senior positions.

Other gender differences in employment conditions also have an impact on occupational safety and health. More women are concentrated in low-paid, precarious work and this affects their working conditions and the risks they are exposed to. Gender inequality both inside and outside the workplace can affect women's occupational safety and health and there are important links between wider discrimination issues and health.

The European Union has stated objectives of further increasing the participation of women and men in employment, improving the quality of working life for all and ensuring equality between women and men. In response to this, the Community strategy on health and safety at work 2002–06 (European Commission, 2002c), drawn up by the European Commission, included 'mainstreaming' or integration of gender into occupational safety and health activities as an objective, recognising the increasing importance of taking account of gender issues in occupational safety and health (OSH).

The current EU approach to occupational safety and health is 'gender neutral', which means that gender issues and differences are ignored in policy, strategies and actions. They are also often ignored in research. OSH policy decisions that appear gender neutral may have a differential impact on women and men, even when such an effect was neither intended nor envisaged. This is because there are substantial differences in the working lives and employment situation of women and men and therefore occupational safety and health, so we need to take account of gender issues in work-related risks and their prevention.

With these EU objectives in mind, this report explores the following questions:

- Are there gender differences in occupational safety and health, such as exposure to hazards, health outcomes and access to resources?
- If so, what are these differences and what causes them?
- What are the gaps in knowledge?
- What is being done to promote occupational safety and health equality in the workplace and mainstreaming gender into OSH?
- What further action can be taken?

The findings of the report include the following key issues:
- Both women and men can face significant risks at work.
- Different jobs, different exposure to hazards.
- Gender segregation in the home: unequal sharing of household duties adds to women's workload.
- Different exposures to work hazards, different health outcomes.
- Reproductive hazards — an unequal focus.
- Examples of hazards and risks in areas of women's work.
- Linking equality and occupational safety and health.
- The risks of ignoring gender.
- Research gaps — improving knowledge of risks to women.
- Promoting equality in prevention: gender mainstreaming and gender impact assessment.
- Taking action to improve gender-sensitivity in risk prevention.

Both women and men can face significant risks at work

Improving the employability of women and men and the quality of working life includes ensuring that: women and men do not have to leave the workforce through injury or ill health; work is compatible with home life; and both can work safely and healthily in a particular job, be it nursing or construction work. There is also a high economic cost of failure to adequately prevent risks at work. The participation of women in the labour market generally appears to be positive for their health. However, both female and male workers can face significant risks from the jobs that they do, underlining the importance of adequate risk-assessment and prevention measures for both. For various reasons, more attention has been paid to the risks that men are more likely to face and their prevention. In contrast, risks to women may be underestimated or ignored altogether.

Different jobs, different exposure to hazards

For all the different types of hazards, both physical and psychosocial, job segregation strongly contributes to work exposure to hazards and therefore to health outcomes. In general, men suffer more accidents and injuries at work than women do, whereas women report more health problems such as upper limb disorders and stress.

Women are more likely to work in jobs involving caring, nurturing and service activities for people, while men are more likely to work in management and manual and technical jobs associated with machinery or plant operation (see Fagan and Burchell, 2002). Even within sectors, there is horizontal segregation, for example in the manufacturing sector women are concentrated in textiles and food processing. This segregation is also vertical, namely men are more likely to work in jobs higher up the occupational hierarchy. For example, while men hold the majority of skilled agricultural jobs, in the lowest level, unskilled occupations, women are disproportionately represented in agricultural-related jobs. Even where men and women appear to be employed to do the same job, in practice, the tasks they carry out can often be segregated by gender.

In addition, women are far more likely to work part-time than men and women part-time workers are even more strongly segregated into female-dominated jobs. Part-time jobs are more monotonous with fewer opportunities for advancement. Women are also more likely to work in the public sector or for small firms and to have temporary contracts (Fagan and Burchell, 2002).

Gender segregation at home: unequal sharing of household duties adds to women's workload

Another striking difference between the working circumstances of women and men is that women still carry out the majority of unpaid work in the home, such as domestic chores and caring for children and relatives, even if they are employed full-time. This adds considerably to their daily work time and puts extra pressure on many women workers. It also increases the costs to the economy of women's work-related illness and injury. It is important to acknowledge that the long and often inflexible working hours of men can also affect the employment and family roles of their female partners.

Influences on gender differences in OSH

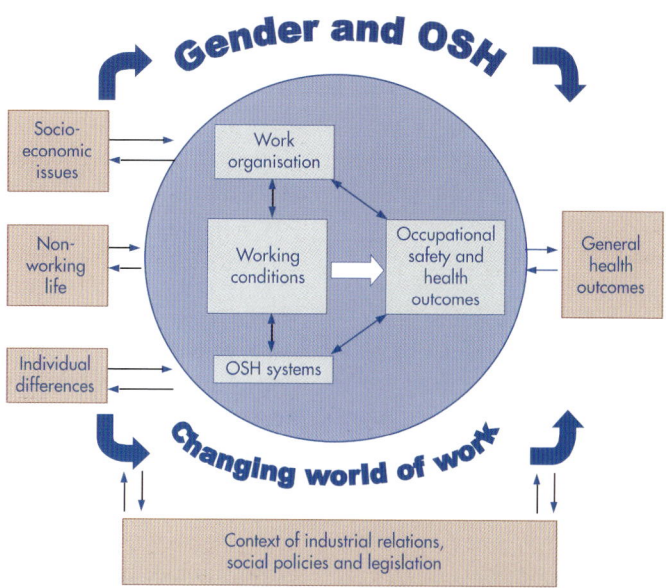

Different exposures to work hazards, different health outcomes

Gender segregation of the workforce strongly contributes to gender differences in working conditions, and hence gender differences in exposures to hazards and health outcomes. In addition, those working part-time will have a lower exposure to hazards than full-time workers in the same jobs. Unless careful account is taken of the gender differences in working conditions, so as to assess real exposure to hazards, it is not possible to assess real rates of work-related accidents and ill health, or ascertain whether women or men are more vulnerable from exposure to certain work hazards.

Even when adjustments are made for the number of hours worked, men still suffer more accidents and injuries at work than women do, whereas women report more upper limb disorders and stress. Occupational cancer is more common among men than women, but there are some occupations, such as food service and certain manufacturing industries, where women have higher rates. Asthma and allergies appear to be more common among women than men. Sources of respiratory hazards in women's work include cleaning agents, sterilising agents and protective gloves containing latex dust used in the healthcare sector and dusts in textile and clothing manufacture.

In addition, women suffer more skin diseases, for example due to working with wet hands in jobs such as catering, or from skin contact with cleaning agents or hairdressing chemicals. Men suffer more from noise-induced hearing loss than women, from exposure to noisy machinery and tools, but women in textile and food production can also be exposed to high noise levels.

Women are more exposed to infectious diseases, particularly in care work but also in the education sector. Men carry out more heavy lifting, but, for example, women in cleaning, catering and care work suffer injuries from heavy lifting and carrying. Women report more upper limb disorders and high incidences are found in some highly repetitive work carried out by women such as 'light' assembly line work and data entry work, where they have little control over the way they work.

Both women and men report high levels of work-related stress; it is certainly not just a 'women's problem'. However, there are certain stressors to which women are more likely to be exposed because of the jobs they typically do. These include emotionally demanding work and work in low-status jobs where they have little control over the work they do. Discrimination and sexual harassment are also sources of stress that women face more than men, as well as the double burden of paid work and unpaid work in the home. Women workers have more contact with members of the public and consequently are more exposed to work-related violence.

Reproductive hazards — an unequal focus

While much attention has been paid to work hazards affecting women who are pregnant or breastfeeding, less attention has been paid to male reproductive hazards, and other reproductive health issues for women, including menstrual problems and the menopause. The European directive to protect pregnant workers should be examined to see how it is operating in practice: are conditions being improved and adjusted in workplaces to protect pregnant workers and is the option to provide extended leave where the risk cannot be otherwise removed really being used only as the final option?

Examples of hazards and risks in areas of women's work

Work area	Risk factors and health problems include:			
	Biological	Physical	Chemical	Psychosocial
Healthcare	Infectious diseases – bloodborne, respiratory, etc.	Manual handling and strenuous postures; ionising radiation	Cleaning, sterilising and disinfecting agents; drugs; anaesthetic gases	'Emotionally demanding work'; shift and night work; violence from clients and the public
Nursery workers	Infectious diseases – particularly respiratory	Manual handling; strenuous postures		'Emotional work'
Cleaning	Infectious diseases; dermatitis	Manual handling, strenuous postures; slips and falls; wet hands	Cleaning agents	Unsocial hours; violence, e.g. if working in isolation or late

Examples of hazards and risks in areas of women's work

Work area	Risk factors and health problems include:			
	Biological	Physical	Chemical	Psychosocial
Food production	Infectious diseases – e.g. animal borne and from mould, spores; organic dusts	Repetitive movements – e.g. in packing jobs or slaughter houses; knife wounds; cold temperatures; noise	Pesticide residues; sterilising agents; sensitising spices and additives	Stress associated with repetitive assembly line work
Catering and restaurant work	Dermatitis	Manual handling; repetitive chopping; cuts from knives; burns; slips and falls; heat; cleaning agents	Passive smoking; cleaning agents	Stress from hectic work, dealing with the public, violence and harassment
Textiles and clothing	Organic dusts	Noise; repetitive movements and awkward postures; needle injuries	Dyes and other chemicals, including formaldehyde in permanent presses and stain removal solvents; dust	Stress associated with repetitive assembly line work
Laundries	Infected linen – e.g. in hospitals	Manual handling and strenuous postures; heat	Dry-cleaning solvents	Stress associated with repetitive and fast pace work
Ceramics sector		Repetitive movements; manual handling	Glazes, lead, silica dust	Stress associated with repetitive assembly line work
'Light' manufacturing		Repetitive movements – e.g. in assembly work; awkward postures; manual handling	Chemicals in microelectronics	Stress associated with repetitive assembly line work
Call centres		Voice problems associated with talking; awkward postures; excessive sitting	Poor indoor air quality	Stress associated with dealing with clients, pace of work and repetitive work
Education	Infectious diseases – e.g. respiratory, measles	Prolonged standing; voice problems	Poor indoor air quality	'Emotionally demanding work', violence
Hairdressing		Strenuous postures, repetitive movements, prolonged standing; wet hands; cuts	Chemical sprays, dyes, etc.	Stress associated with dealing with clients; fast paced work
Clerical work		Repetitive movements, awkward postures, back pain from sitting	Poor indoor air quality; photocopier fumes	Stress, e.g. associated with lack of control over work, frequent interruptions, monotonous work
Agriculture	Infectious diseases – e.g. animal borne and from mould, spores, organic dusts	Manual handling, strenuous postures; unsuitable work equipment and protective clothing; hot, cold, wet conditions	Pesticides	

Linking equality and occupational safety and health

There are important links between wider issues of discrimination and women's work-related health, which were largely outside the scope of the study. However, gender differences in social conditions and employment conditions have an impact on occupational safety and health and cannot be ignored. As has been mentioned, more women are concentrated in low-paid, precarious work and this affects their working conditions and the risks they are exposed to. Issues such as sexual harassment and discrimination in the workplace are two stress factors that women face more than men. As gender inequality both in and outside the workplace can have an impact on women's occupational safety and health, it should be included or 'mainstreamed' into equality agendas. Women's weaker participation in all levels of occupational safety and health consultation and decision-making will also contribute to less attention being paid to their OSH needs, and poorer risk assessment, if they are not effectively consulted (see below).

The risks of ignoring gender

There is evidence that taking a gender-neutral approach to occupational safety and health is contributing to the maintenance of gaps in knowledge and to less effective prevention. For example, it contributes to:
- gender differences being obscured or overlooked;
- less attention being paid to some research areas more related to women;
- the extent of risks to women being underestimated;
- reduced participation of women in occupational safety and health decision-making, as positive action is not taken to ensure that they are included;
- not the most appropriate preventive solutions being selected.

Policy decisions on priorities for legislative action or labour inspectorate support and enforcement activities are increasingly based on risk analysis. Therefore, to ensure equality of treatment, it is very important that risks to women are not underestimated. On the other hand, where gender has been taken into account, the focus has been on women as childbearers, which has led to a neglect of other reproductive hazards to both men and women.

Research gaps exist — improving knowledge of risks to women

More attention is now being paid to the occupational safety and health of women workers. However, work-related health and safety risks to women can be underestimated in various ways. Methods to obtain more accurate information include the following:
- Systematically and routinely building gender questions and analysis into occupational safety and health monitoring and research.
- Adjusting data for hours worked (women generally work fewer than men) and breaking them down more accurately according to actual jobs performed.
- Looking at figures related to specific jobs, rather than average figures.
- Including both genders in research, and carrying out analyses so gender differences can be examined, rather than 'controlling' for gender. For example, coronary heart diseases are a major cause of death for women, but despite this occupational research has concentrated largely on men. This is partly due to the fact that women develop heart diseases later than men, when they have retired, so the possible occupational connection has often been ignored.
- Ensuring that work-related injuries and ill health that are relevant to women workers are covered in statistical monitoring and research. For example, injuries from work-related violence are not always included in

national statistics and work–life balance issues have not always been included in workplace stress investigation.
- Paying attention to previously ignored women's health issues in the occupational setting. For example, possible links between some women's health issues, among them menstrual disorders and the menopause, and occupation.
- Improving data collection on the links between women's ill health and occupation. For example, by recording women's occupation in epidemiological health monitoring and in death certificates, etc.
- In research, looking at the real jobs done, not simple job titles and descriptions, and the related risk exposure, and taking a worker participation approach.
- Targeting research into risks to women workers and their prevention. For example, the US Occupational Safety and Health Institute, NIOSH, introduced a research programme targeting women and female-dominated jobs.

In addition, existing epidemiological research should be critically assessed to find any systematic bias in the way investigations are carried out when studying women and men's health and illness patterns in order to develop improved methods.

Promoting equality in prevention: gender mainstreaming and gender impact assessment

There is a variety of EU legislation, or directives, on gender equality, but the approach of EU directives on occupational safety and health is generally gender neutral. This means that the OSH approach does not specifically take account of gender issues. However, we find that there is better coverage in specific EU directives of risks that men are more commonly exposed to, such as noise, or male-dominated work areas, such as construction, than risks that women are more commonly exposed to, such as upper limb disorders and stress. One area of female-dominated work, paid domestic work, is excluded from coverage by the OSH directives altogether. Many occupational safety and health standards and exposure limits to hazardous substances are based on male populations or laboratory tests and relate more to male work areas. In addition, listed occupational accidents and diseases for compensation purposes provide better coverage for work-related accidents and ill-health problems that are more common among men. These issues need to be addressed.

The Community strategy on safety and health at work 2002–06 includes gender mainstreaming as a key objective, for example by incorporating gender elements into risk-assessment activities, implementation of preventive measures, compensation arrangements and benchmarking actions. Mainstreaming concerns the integration of gender issues into the analyses, formulation and monitoring of policies, programmes and projects to reduce inequalities between men and women. This implies that gender should be taken into account throughout the implementation of the Community strategy, including areas of development and promotion of political instruments, such as legislation, social dialogue, progressive measures and best practices, corporate social responsibility, economic incentives and mainstreaming occupational safety and health into other policy areas.

The findings of this report support the need for gender mainstreaming, and gender impact assessment should be used as one of the tools to achieve this. Gender impact assessment, which forms part of the Commission's gender equality strategy, should be applied to all areas of occupational safety and health policy, including reviews of existing occupational safety and health directives, development of new legislation and guidelines, the standardisation process and compensation arrangements.

Despite the shortcomings of a gender-neutral approach in EU occupational safety and health legislation and guidelines, and the need to subject them to gender impact assessments, it is possible to apply them in a gender-sensitive way. If effectively implemented and enforced in all Member States, they could bring significant improvements to many women's jobs. Guidelines, risk-assessment instruments and training are necessary to support the gender-sensitive application of the directives.

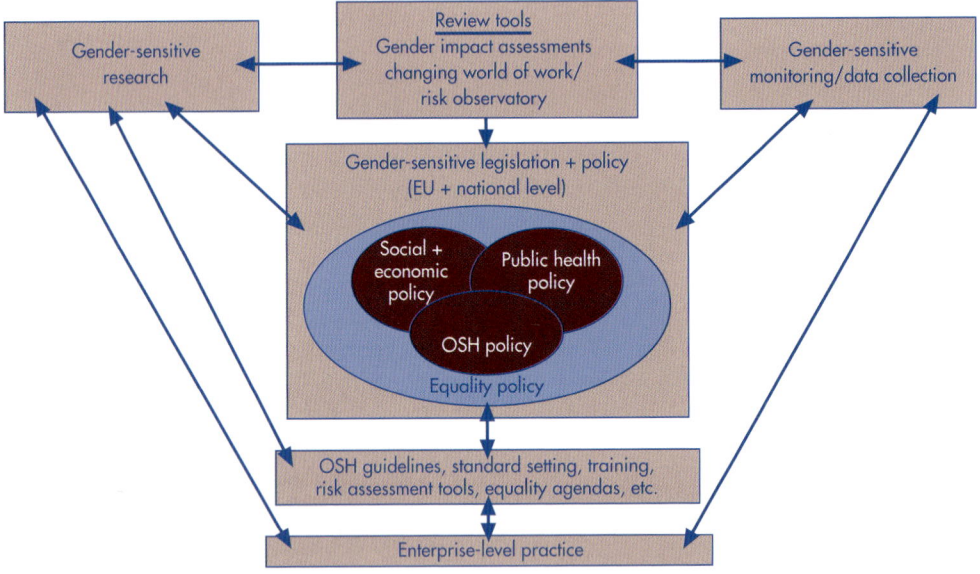

Action levels for mainstreaming gender into the OSH system

Taking action to improve gender-sensitivity in risk prevention

Gender-sensitivity relates to taking account of gender issues, differences and inequalities in strategies and actions. The findings of this report regarding existing research and good practice, and recommendations of bodies such as the International Labour Organisation (ILO) and World Health Organisation (WHO) and by contributors to a Trade Union Technical Bureau (TUTB) report (Messing, 1999) suggest that there are already various actions that can be taken towards achieving a more gender-sensitive approach to risk prevention in EU workplaces.

Workplace level — enterprises

Although there are some gaps in what we know, there is sufficient scientific knowledge to allow us to take action to prevent the work-related risks to women covered in this report. What is particularly important is that, during risk assessment, the jobs of both women and men are considered, and that this includes looking at the jobs/tasks that they really do in practice and all the influences on them. This should take place with the effective participation of all in the enterprise. At the workplace level, the following steps should be taken:

- Take a gender-sensitive approach to risk assessment, and ensure that information and training on gender issues in OSH are provided.
- Make links between equality activities and OSH activities: for example, include working conditions in equality policies and actions, and include issues such as sexual harassment and work–life balance in OSH activities, for example in stress prevention. Look at issues such as flexibility and enrichment of jobs, and working schedules of both women and men in order to improve work–life balance.
- Facilitate the participation of women in OSH consultation, decision-making activities and safety committees.
- Take account of both OSH and the gender impact of any changes in work organisation.
- Use the risk-assessment model and practical examples given in this report as a starting point for action, adapting them as necessary.

Social partners

- Take account of gender in all OSH activities.
- Use social dialogue committees, at EU or national level, to develop guidelines and action plans and promote a more gender-sensitive approach.
- Raise awareness of gender issues in OSH.
- Develop guides and training resources and encourage more women to be involved as worker representatives or on safety committees, etc.
- Include OSH in equality activities.

OSH authorities

- Ensure that occupational safety and health policies and programmes contain a well-defined and transparent gender dimension.
- Develop guidelines and inspection tools, for example for applying regulations in a gender-sensitive way, and training resources.
- Provide training and awareness raising for labour inspectors and for workplaces.
- Ensure that resources and intervention activities are directed towards jobs of both women and men.
- Promote research into risks or jobs of particular relevance to women, where these have been neglected.

Policy level

- Apply gender impact assessments to OSH directives, legislation, limits and standards, benchmarking, priority setting, compensation arrangements, etc.
- Develop new OSH policies in a gender-sensitive way.
- Set up advisory groups on mainstreaming gender into OSH.
- Include gender in all OSH activities.
- Include OSH in equalities policy activities.
- Promote the involvement of women in OSH policy setting.
- Include gender issues and risks to women workers in OSH research programmes.
- Promote a holistic approach to OSH that covers the work–home interface.
- Promote interdisciplinary cooperation.

Key conclusions

- Continuous efforts are needed to improve the working conditions of both women and men. Making jobs easier for women will make them easier for men too.
- There is strong segregation of women and men into different jobs and tasks at work, and women carry out a greater proportion of unpaid duties in the home. Women are also more likely to work part-time. This gender segregation at work and home and gender differences in employment conditions have a major impact on gender differences in work-related health outcomes. Research and interventions must take account of the real jobs that men and women do. More studies are needed that examine women and men carrying out the same tasks under the same circumstances

and take account of differences in exposure and working conditions.
- In addition to basing exposure assessment on the real jobs people do, research and monitoring can be improved by systematically including the gender dimension in data collection, and adjusting for hours worked. Epidemiological methods should be assessed for any bias. Indicators in monitoring systems, such as national accident reporting and surveys, should effectively cover occupational risks to women.
- Work-related risks to women's safety and health have probably been underestimated and neglected compared to men's. Various risk factors in women's work and their interactions have been insufficiently examined, and have received less attention in risk-assessment and preventive activities. Neglected areas should be specifically prioritised for research, awareness-raising and prevention activities.
- Taking a gender-neutral approach in policy and legislation has contributed to less attention and fewer resources being directed towards work-related risks to women and their prevention. In addition, domestic workers are specifically excluded from the coverage of the directives. Women working informally, for example wives or partners in family farming or fishing businesses, may not always be covered by legislation. Gender impact assessments should be carried out on existing and future OSH directives, standard setting and compensation arrangements.
- Although gender impact assessments of OSH legislation are necessary, enough is known about both prevention and mainstreaming gender into OSH to enable current directives to be implemented in a more gender-sensitive way.
- For success, gender-sensitive interventions should take a participatory approach, involving the workers concerned, and be based on an examination of the real work situations.
- Improving women's occupational safety and health and quality of working life cannot be viewed separately from wider discrimination issues at work and in society. It is essential that OSH be integrated into employment equality actions and activities to mainstream occupational safety and health into other policy areas, such as public health or corporate social responsibility initiatives, should include a gender element.
- Women are under-represented in decision-making concerning occupational health and safety. They need to be more directly involved and women's views, experiences, knowledge and skills should be reflected in formulating and implementing health-promotion strategies within OSH.
- There are already successful examples of including or targeting gender in research approaches, interventions, consultation and decision-making, tools and actions. Existing experiences, resources and good practices in mainstreaming and taking a gender-sensitive approach should be shared and exchanged.
- While the general trends in women's working conditions and situation are similar across the Member States and candidate countries, there are also country differences within these general trends. It is therefore important that individual countries examine their particular circumstances regarding gender and OSH, in order to plan actions.
- Taking a holistic approach to OSH, including the work–life interface, broader work organisation and employment issues and all issues related to the changing world of work, would improve occupational risk prevention for the benefit of both women and men.
- Women are not a homogeneous group, they fall into different age groups, have different ethnic origins and such like and not all women work in traditionally 'female' jobs. The same applies to men. A holistic approach will need to take account of diversity. Actions to improve work–life balance must take account of both women's and men's working schedules and be designed to be attractive to both.

1. INTRODUCTION

Why study health and safety at work from a gender perspective?

Women's participation in the labour market has been rising and the Lisbon European Council (23 and 24 March 2000) set itself the task of achieving an overall activity rate in the labour market of 70 %, and 60 % for women. Providing a safe and healthy workplace is an important part of ensuring and maintaining employability, together with reconciliation of working and non-working life. Prevention of ill health and the promotion of well-being at work are important for the quality of work of both women and men and the costs of work-related ill health are high for employers and society as well as the individuals concerned.

The promotion of equality is an important area of European Community social policy and this includes equality in working conditions. Integrating gender into Community policy areas is what is termed 'gender mainstreaming'. The gender issue was included in the Community strategy for safety and health at work for 2002–06 (European Commission, 2002c). The strategy notes the increasing participation of women in employment, and acknowledges gender differences in the incidence of occupational accidents and diseases. The strategy puts forward a number of objectives for a holistic approach to well-being at work that must be targeted jointly by all players. These include 'mainstreaming the gender dimension into risk evaluation, preventive measures and compensation arrangements, so as to take account of the specific characteristics of women in terms of health and safety at work'. The European Parliament, in its comments on the occupational safety and health strategy, strongly emphasised the importance of the gender element and highlighted a number of areas for attention (see Annexes 3 and 4).

In order to take the strategy on mainstreaming gender into occupational safety and health forward, it is important to look critically at what we know about any gender differences in safety and health at work, why there could be differences and any gaps in knowledge. With this in mind, the Administrative Board of the European Agency for Safety and Health at Work included a review of gender and occupational safety and health in the Agency's 2002 programme of work.

There are various other developments that make it important to look at gender differences in occupational safety and health. Women have entered the service sector in particular and they are over-represented in part-time and short-term contract work. How does this affect their occupational safety and health? In addition, are there gender differences in occupational safety and health between the EU Member States and the candidate countries (¹)?

Statistics show that men are more likely than women to suffer accidents at work, including fatalities, but they also show that work can be a serious source of ill health and work absenteeism for women.

There is growing awareness that women and men are exposed to different hazards at work and have different health outcomes. Nevertheless, women's risks in the workplace are still often overlooked. Both research and prevention activities have traditionally targeted risks in jobs dominated by men.

Key questions to be addressed

Do the bare statistics tell the whole story? To what extent are any differences in women's and men's work-related health due to sex differences (biologically determined differences) and to what extent gender differences (social differences that are changeable and variable)? How do the different jobs and tasks women and men carry out affect their occupational safety and health? Are there additional risks for women entering male-dominated employment areas? How are women and men provided for by occupational health services? It is assumed that a safe and healthy workplace will suit everyone the same. What effect does this gender-neutral approach have? What effect do women's greater responsibilities in the home have? Is the same picture seen in the Member States and the candidate countries? What do the answers to these questions mean for risk prevention?

This report looks into the questions listed above and other issues. The key questions addressed are as follows:
- Are there differences between men and women in occupational safety and health, and, if so, what causes these differences?
- How gender-sensitive is current research and what are the gaps in knowledge?
- What has been done and what can be done to promote occupational safety and health equality in the workplace?
- How gender-sensitive is the current occupational safety and health system and what are the needs for action?
- How can gender mainstreaming be incorporated into legislation and into occupational safety and health management?

What to include — the conceptual model

Many factors inside and outside the workplace can influence the health and well-being of workers. These include physical working conditions, the occupational safety and health system, life outside work and social policy. It is important to consider all these factors and how they interact when looking at gender differences in occupational safety and health. The following model seeks to show these interrelated factors. However, it has not been possible to explore all the issues in depth in this report and the main focus is on the middle section of the model, occupational safety and health in the workplace (see Figures 1 and 2).

The report starts by looking at differences in the general employment position between women and men. It then covers a number of specific hazards, jobs and health outcomes. This section also pays attention to specific groups, such as older workers. Then, the context of risk identification and prevention is considered in more detail, including information, occupational health

---

(¹) Candidate countries refer to 13 candidates for entry into the European Union at the time of writing the report: Bulgaria, Cyprus, the Czech Republic, Estonia, Hungary, Latvia, Lithuania, Malta, Poland, Romania, Slovakia, Slovenia and Turkey. This report refers to the European Foundation's survey of working conditions in 12 of these candidate countries, as the survey did not cover Turkey (Paoli, 2002).

Figure 1: Theoretical framework for the study

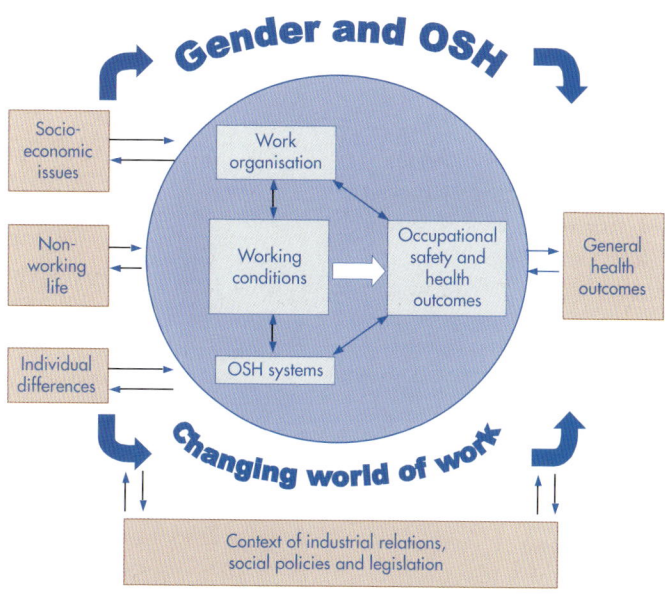

Figure 2: Detail of elements included in the theoretical framework

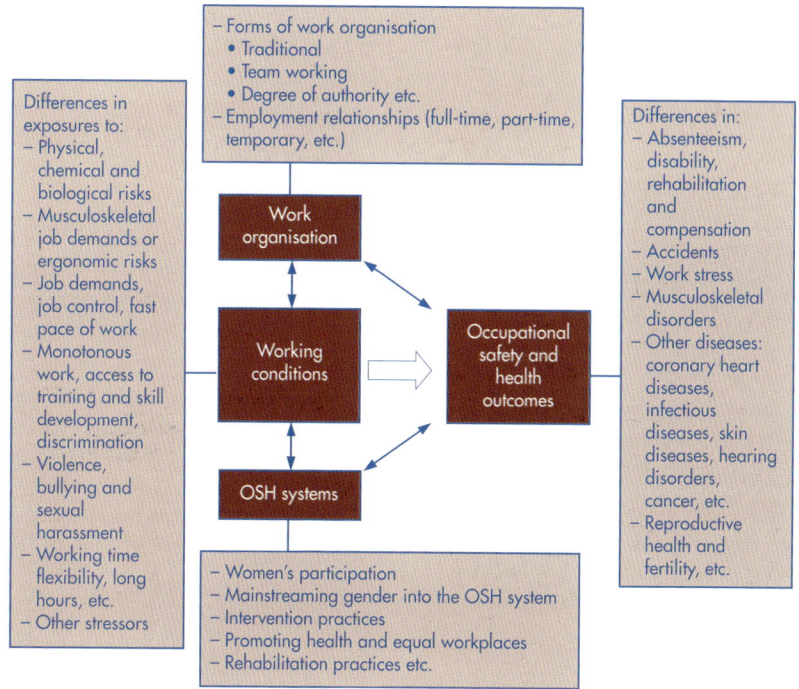

systems, participation of the workforce and mainstreaming. The report presents conclusions and suggestions for taking forward the goal of equal health at work.

The methodology

The European Agency's Topic Centre on Research wrote the original manuscript that was used for the basis of this report. Information was gathered through a prior survey and discussion at a meeting of experts, a literature survey and Internet searches. The theoretical framework for the report was also developed at the expert meeting. The Agency's network in the Member States and European Free Trade Association (EFTA) countries and a further group of experts in this area commented on the original manuscript. The final report was edited by the Agency, in cooperation with the topic centre. Details of all those involved are given in Annexes 1 and 2.

Concepts and terminology

Definitions of terms such as 'gender', 'gender-sensitive' and 'gender mainstreaming' are given in the glossary.

A word on statistics

This report draws heavily on statistics from three main sources: official records of accidents from the Member States, which are collated and analysed by Europe's statistical office, Eurostat; a self-report labour force survey carried out in the individual Member States, the results again analysed by Eurostat; and the European survey on working conditions carried out by the European Foundation for the Improvement of Living and Working Conditions. These statistics are supplemented by information from individual Member State and research studies. Each source has its limitations, both in general and for obtaining an accurate picture of gender differences.

National statistics from the Member States are not directly comparable, as not all collect the same information in the same way. Only information on serious accidents is collected, such as accidents resulting in an absence from work of more than three days. We will see that women are more likely to suffer work-related diseases and men accidents. The recorded Member State information on cases of work-related diseases is quite limited and even more variable. We will see that workplace violence is one area that affects women. Not all Member States collect information on cases of work-related violence, and not all countries collect information by gender.

The self-report surveys have been carried out to obtain information on the occurrence of less serious accidents and more information on work-related diseases. Although they give a broader picture, these too have their drawbacks. They are only as good as the questions asked. The replies are subjective and answers may depend on awareness and understanding of the subject matter of the question — for example, it has been assumed that wide country variations in response to a question in the European survey on working conditions about harassment at work are probably due to differences in awareness levels. Not all Member States took part in the labour force survey. Ideally, to ensure that the differences are not masked and that the respondents understand questions correctly, survey studies should be complemented with discussions in focus groups, for example. Fagan and Burchell (2002) have made a number of recommendations for improving the gender-sensitivity of the European survey on working conditions, in order to get a clearer picture of gender differences and exposure to hazards that particularly affect women. For example, improvements are suggested to the current question on violence, harassment and intimidation.

We shall see that the jobs people do and where they work have a great influence on the hazards that they are exposed to. However, currently, only rather broad information about sector and job is collected. Industrial structure and sectoral distribution of the workforce, including with regard to gender, differ in each country too. The likelihood of having an accident or suffering a work-related disease is greatly determined by level of exposure. Women work fewer hours than men, and hours worked also vary by sector. So, when looking at gender differences in occupational accident and ill-health rates, it is very important to adjust the figures for the number of hours worked. Eurostat has carried out some secondary analyses of the data it collects, and these statistics are used in this report (Dupré, 2002). The European Foundation has also made an extensive analysis of its survey information by gender, including analyses by gender and full-time and part-time working, and we draw considerably too on these results in this report.

2.

GENDER DIFFERENCES IN WORKING LIFE IN EUROPE

Key points

- Women's participation in the workforce has increased. They now make up 42 % of the workforce in the EU.
- Women are far more likely to work part-time than men. Men are more likely to work long hours.
- The labour market is heavily segregated horizontally. For example, there are more women in the public sector, and in the service sector and sales and who work as office clerks. Men are more likely to be machine operators, hold technical jobs, and to work in craft trades, construction, transport and extraction industries.
- Even within the same jobs in the same organisation, women and men often carry out different tasks.
- The labour market is also heavily segregated vertically, with more men in managerial and senior positions.
- Women are more likely to work in 'people' work, including caring, 'nurturing' and service work. Men dominate management and jobs associated with machinery or physical products that are considered to be 'heavy' or 'complex'.
- In the elementary occupations, women are more likely to work in cleaning and agriculture and men, in general, in labouring.
- Women are under-represented in permanent contracts and among the self-employed.
- Men earn more than women, even after figures are adjusted for number of hours worked.
- Outside of work, working women are more likely to take care of their children, take care of elderly or disabled relatives, and do the cooking and the housework than working men.
- Compared to the EU, in the candidate countries, there is less gender segregation, working hours are less gender differentiated, there is less part-time working, and the dual workload of paid work and unpaid household/caring is more gender balanced.

Introduction

Gender differences and work-related risks cannot be discussed without looking at the position of women at work and the differences men and women have in their working lives. For example, exposure to hazards varies according to number of hours worked and between different sectors, jobs and tasks because the working conditions are different. It is also important to look at assumptions about the 'riskiness' of different jobs women and men do and where the focus of attention for occupational safety and health has been.

Women's participation in the labour market

Women's share of total employment in the European labour market has grown considerably over the last 25 years. However, women still have a lower participation rate than men and generally work shorter hours. There is considerable variation in participation across the Member States. For example, women's participation is greatest in Sweden where women's share of total employment reached 48 % in 2000, while the figure for Spain is 35 % (see Figure 1 in Annex 5). A higher proportion of women work in the candidate countries, 46 % compared to 42 % in the Member States, according to European Foundation surveys on working conditions (Paoli, 2002).

There are also great differences between the participation rates of women belonging to different age groups. The participation rate of women aged 25–49 is highest in Scandinavian countries (about 80 %) and lowest in Italy, Spain and Greece (about 55 %). Scandinavian countries also have the highest participation rate among women aged 50–64, as about 60–70 % of women belonging to this age group work outside their home; in Greece, Luxembourg, Belgium, Spain and Italy, only 20–30 % of women of this age group work (see Figures 2 and 3 in Annex 5). Women are less likely than men to work in each age band, but the difference is most marked in the middle years. It is important to pay attention to age structure, as young and old workers may have different occupational safety and health outcomes.

Having children strongly influences women's participation rate. In 2000, 72 % of women aged 20–50 without children were in work, compared to only 59 % of women with children under six. As opposed to this, 94 % of men aged 20–50 with children were working, compared to 89 % with children (European Commission, *Social Agenda*, July 2002). Participation rate and number of hours worked are influenced by the availability of childcare and other measures to facilitate the combination of employment and family life (European Foundation, 2002c). Many women returnees enter low-skilled, low-status jobs, such as cleaning or domestic work (McCarthy and Scannell, 2002).

The main reasons for European women's inactivity in employment are lack of education and training (27 %), personal or family responsibilities (20 %), early retirement (16 %) and illness or disability (9 %) (European Commission, 2002d; report requested by the Stockholm Summit).

Working time

The main findings from the third European survey on working conditions regarding working time and gender are given in Box 1. When employed, women generally work shorter hours than men. Some 20 % of employed men and 8 % of women regularly work long days of 10 hours (Fagan et al., 2001; Fagan and Burchell, 2002). The main growth in employment in the EU has been in part-time jobs (Clarke, 2001). The employment participation rate of women, in particular, has increased through entry into part-time jobs and there is a significant gender difference in the number of hours worked. According to the third European survey on working conditions, while 17 % of respondents worked part-time, the figure was 32 % for women and only 7 % for men. Again, large country differences are seen, with 33 % of people reporting working part-time in the Netherlands and only 5 % in Greece (see also Figure 4 in Annex 5).

Differences between full-time and part-time workers can also be seen in the type of work carried out, working conditions and training provision. Part-time work is more typical of the service sector, sales jobs, clerical work, and the lowest level, unskilled manual jobs. Senior posts are rarely available part-time.

> **Box 1: Gender differences in working time — some findings from the third European survey on working conditions**
>
> - Women's weekly working hours are in general nine hours less than men's.
> - 20 % of employed men and 10 % of employed women work very long weekly hours (48+).
> - Women working in the male-dominated areas of management and blue-collar craft jobs are more likely to work long days than in other occupations, but in management frequent long days are still less likely for women than men.
> - 32 % of women and 7 % of men report working part-time.
> - Men are more likely than women to have some possibility to vary their start and finish times (26 % men compared to 19 % women) and this gender difference is most pronounced in white-collar work among professionals and managers.
> - 20 % of employees are shift-workers — the proportion is similar for full-timers and part-timers and by gender.
> - Among professionals, women are more likely to work rotating shifts than men. An example is healthcare workers. Among those in blue-collar operating jobs, men are more likely to work rotating shifts.
> - Women shift-workers are more likely to work nights if they are employed part-time.
> - Full-timers, particularly men, are likely to have the longest commuting times.
> - Both women and men working full-time are more likely to say that the demands of their job are not compatible with family and social commitments and the lack of fit is more acute for parents; a quarter of all men and a quarter of women working full-time report 'work–life' incompatibility.
> - Half of mothers working part-time report that their hours are compatible with other commitments, meaning that the other half do not have such a good fit.
> - Men are more likely than women to report that working 'non-standard' work schedules causes incompatibility with life outside work, often because this is combined with long hours.
> - Differences are seen by age and country.
> - Differences are shaped by domestic situations, labour market opportunities, employment area, legislation and social policy, etc.
> - Working hours are longer and tend to be less gender differentiated in the candidate countries compared to the EU. This reflects the fact that the level of female part-time work is low. For both men and women, part-time work is less frequent — 7 % of employed people report working part-time in the candidate countries compared to 17 % in the EU (additional information from a survey of working conditions in 12 candidate countries).
>
> *Sources:* Paoli and Merllié (2001); Fagan and Burchell (2002); Paoli (2002).

The number of hours worked affects the amount of exposure to work hazards. However, analysing the data from the third European survey on working conditions, Fagan and Burchell (2002) found a number of differences between full- and part-time workers, including the following:
- Part-time jobs offer fewer opportunities for learning and have fewer planning responsibilities. Women part-timers receive the fewest.
- Part-timers are most likely to say that their skills are underused, and they report receiving less training than full-timers. Women part-timers report receiving the least training. (See also the section on employment contracts.)

Factors such as underuse of skills can contribute to stress. Lack of training can mean you do not have the training to do your job properly or safely and it can also hinder career advancement.

Working time will also be discussed again in the section dealing with work–life balance.

Employment status

Contrary to popular opinion, salaried employment, as opposed to self-employment, is still the norm (82 %). However, the growth in temporary work has resulted in a greater incidence of fixed-term and temporary agency work (European Foundation, 2002a). Some gender differences in employment status are given below (see also Figures 5 and 6 in Annex 5).

According to respondents to the third European survey on working conditions:

- the majority of both working men and women have a standard contract of unspecified duration;
- the rate of self-employment is 19 % for men and 12 % for women;
- women are slightly more exposed to the insecurity of fixed-term contracts;
- part-time workers of either sex are more likely to be in the precarious position of holding a temporary contract than full-timers;
- 32 % of employed women and 7 % of employed men report working part-time;
- there are differences among the Member States.

Sources: Paoli and Merllié (2001); Fagan and Burchell (2002).

There are country differences (see Figure 5 in Annex 5). Spain has the highest number of non-permanent contracts in the EU countries and, excluding the UK and France, more women than men work on a non-permanent basis. Non-permanent employment contracts include seasonal contracts, temporary agency contracts, and work on call. In Finland, non-permanent contracts are typical of the female-dominated sectors of healthcare, social work and teaching (Sutela et al., 2001). Naumanen (2002) found that in Finland young women with a university degree were much more likely to work on a non-permanent contract than men.

There are also age differences (see Figure 6 in Annex 5). The greatest differences in employment status between men and women are due to women working part-time, and women between 25 and 45 years of age working on non-permanent employment contracts. The majority of men will only work on a non-permanent contract at the start of their career. Over 90 % of all male employees of 25 years or older work on a permanent contract (mainly full-time). Women are more likely to keep a precarious position in the labour market throughout their working lives: a higher percentage of women than men above 25 years of age have a non-permanent contract: 15 % of women in the age group 25–39 have a non-permanent contract (part-time or full-time) and 8 % of women in the age group 40–45. Approximately 35–40 % of female employees have a permanent part-time contract.

Gender differences in the length and type of employment contract are probably to be found in the simultaneous growth of the service sector, growth in the number of non-permanent contracts and entry of more women into the labour market in the past decade. In addition, women's greater role in childcare will cause them to step in and out of the labour market more often than men, and seek part-time jobs or on-call contracts.

Implications for occupational safety and health

Studies by the European Foundation have found a clear correlation between precarious forms of employment and health (European Foundation, 2002a). For example, Benach et al. (2002) studied the health impacts of different types of employment based on the data collected by the European survey on working conditions. When compared to permanent workers, non-permanent workers reported more job dissatisfaction. Full-time employees with a fixed-term contract reported high levels of muscular pains and fatigue, compared to full-time employees with a permanent contract. Benach and colleagues used gender as one of the control variables and concluded that both the association between non-permanent employment and dissatisfaction continued after taking into account individual variables such as gender.

On the basis of the same data, conclusions that can be drawn about the working conditions of non-permanent employees include (see Goudswaard et al. 2002; Goudswaard and Andries, 2002): non-permanent employees face worse ergonomic conditions than permanent employees, but these differences can be explained by different sectors and occupations. Both non-permanent employees and part-timers have less job security, less job control, less access to training, less skills development and poorer career prospects than permanent employees. Looking at job control as a way of coping with high demands, it may be that the consequences of intensification can be experienced more by non-permanent employees, since they have less job control.

As has been mentioned, part-time employees may also have less access to training. In part-time service work, work may be concentrated in peak hours and therefore be more intense. Since women do more part-time work and have more non-permanent contracts than men, this could influence any gender differences seen. However, analysis is needed by employment status, occupation and gender to be able to draw conclusions about what is causing any differences seen.

Box 2

Compared to full-time jobs, part-time jobs are:

- lower paid, including often in terms of hourly rate;
- segregated into a narrower range of occupations, particularly in the service sector;
- typically more monotonous with fewer opportunities for learning or formal training;
- less likely to have an intense pace of work.

Compared to full-time workers, part-time workers are:

- less exposed to hazards, being protected by working fewer hours;
- more satisfied with working hours and less satisfied with pay, access to training and promotion opportunities.

Source: Fagan and Burchell (2002).

Gender differences in unpaid household and caring tasks

At the beginning of the report, a model was presented to show that some of the gender differences in working conditions are related to the broader pattern of gender inequality in society. One important such area is women's double shift of paid and unpaid work, as they do more household chores and care work in the home (for example, see Fagan and Burchell, 2002). Unpaid duties can include cooking, shopping, housework, and caring for children and elderly relatives.

In the third European survey on working conditions, 63 % of working women and 12 % of working men reported doing housework for one hour or more each day. In all, 41 % of women and 24 % of men reported taking care of children and their education for one hour or more each day. Also, 6 % of women compared to 2 % of men reported involvement in the care of elderly and disabled relatives for one hour or more each day (European Foundation, 2000c; Fagan and Burchell, 2002. See also Table 9 in Annex 6).

The dual workload is more balanced between the sexes in the candidate countries, although it is still far from being evenly distributed (Paoli, 2002). However, one reported impact of transition in some candidate countries, such as Lithuania, has been the removal of pre-school childcare in workplaces (Gonäs, 2002).

Time for activities outside of work and home duties is important for quality of life. According to the third European survey on working conditions, women reported having less engagement in leisure, sport, cultural activities or even sleep than men, particularly when they were mothers, living with a partner and employed full-time. As opposed to this, single and childless women are more likely to be involved in education and training activities than men (Fagan and Burchell, 2002. See also Table 10 in Annex 6).

On average, employed women spend more time in total than men on paid and unpaid work, particularly if they work full-time. Although there has been an increase in the amount of time men devote to household tasks and caring, it is clear that the main burden still falls on women. We will return to this issue later, and look more into the topic of work–life balance.

The impact of changes in the world of work

Changes in the world of work are taking place, influenced by technological developments, changing market pressures, globalisation, economic policies, etc. These changes include more flexible contracts; growth in small and medium-sized enterprises; leaner organisations with decentralised management structures; greater use of subcontractors; new forms of work organisation; network organisations; less stable and more diffuse borders between companies; more individualised relationships within companies; and less precise boundaries between work and private life (European Agency, 2002f; Goudswaard et al., 2002; see also Castells, 1996; Ekstedt, 1999; Sennet, 1998). The rapidity of change and the complexity of work organisation also pose challenges for the effective management of occupational safety and health. For example, due to the increasing interdependency between companies, cooperation on matters concerning work-related health becomes crucial. Various reports have therefore concluded that it is necessary to study the occupational safety and health impact of the changes that have been taking place in the world of work, for example, Cooper (1998); European Agency (2002f); Goudswaard et al. (2002). Within this, more attention should be given to the gender dimension (Goudswaard et al., 2002).

Work intensification

The results of the European surveys on working conditions suggest an intensification of work in all EU countries and all sectors of industry, with work being carried out faster. The changes in enterprises and work organisation, including market pressures, may be one of the reasons for this (European Foundation, 2002a). Looking at gender differences in the third European survey, women are more likely to have their pace of work set by the demands of people, and men are more exposed to production targets. No gender difference was found in the requirement to work at high speed or whether respondents had sufficient time to do the job (Fagan and Burchell, 2002). This is in line with other literature, for example Vinke and Wevers (1999) and Goudswaard et al. (1999). Intense work is directly related to both stress and musculoskeletal disorders (Buckle and Devereux, 1999; Cox et al., 2000; European Foundation, 2002a). There may also be a link between workplace bullying and time pressures in the workplace to get things done.

Market and client-driven work — the flexible firm

Companies have sought to become more responsive to clients and are putting a greater emphasis on being flexible to market changes.

This, together with the greater use of information technology, has had an influence on jobs and working conditions.

Vickery and Wurzburg (1996) claim that about a quarter of companies can be described as 'flexible firms'. Oeij and Wiezer (2002) estimate the prevalence of new forms of work organisation to be between 10 and 25 %.

Two types of resultant 'flexible' organisations have been described. Firstly, there are organisations with a decentralised management and with a low level of operator autonomy — so-called 'lean' organisations. This first type is characterised more by stressful working conditions due to the combination of high job demands, low autonomy and lack of opportunity to learn on the job. Secondly, there are organisations with decentralised management, but with a relatively high level of worker autonomy, for example using autonomous groups. The existence of more autonomy and requirement of more skills and the possibility to develop skills mean that this type of organisation is noted for less stressful working conditions. Therefore, not all 'new forms of working organisation' are contributing to improvements in working conditions (Daubas-Letourneux and Thébaud-Mony, 2002; European Foundation, 2002a).

Oeij and Wiezer (2002) observed feelings of job insecurity due to changes in firm structure and strategy, threats to health and well-being as a consequence of psychosocial and physical risks, which are related to increased competition and labour productivity, an ongoing pressure on competence development, and possible problems in attuning work to private life. They suggest that women will experience more job insecurity than men due to the fact that they work more on non-permanent contracts than men. There is, however, no evidence of such a relationship and this issue needs to be studied further.

Organisational change potentially provides an opportunity to address issues such as gender segregation and inequalities in working conditions, and direct participation of workers in the change process, with direct worker participation being an important part of the change process. This does not appear to be happening. Webster and Schnabel (1999) suggest that direct participation does not result in more attention being paid to these areas either, although organisational change could be used to introduce mixed-sex team working or multi-skilling of men and women. More research is needed into the mechanisms that could be used to reduce gender segregation and inequality in working conditions during organisational change.

Box 3: Call centres — a new area of work

Call centres are a growing industry in Ireland and elsewhere in Europe. It is estimated that more than a million Europeans will be working in call centres by the end of 2003. The aim for employers of call centres is to increase service efficiency and cost efficiency. The great majority of the employees in call centres are women. Employee dissatisfaction is reported to be high, which is manifested as a high agent turnover. Lack of career advancement, the repetitive nature of the work, tightly scripted work, mentally and physically demanding work, difficult customers and 'being in a situation where one is not in a position to make a decision' are among the most common complaints of the employees. The result is frustration, emotional exhaustion and disillusionment. In addition, feeling stressed and uncommitted is common (McCarthy and Scannell 2002).

Although psychosocial risk factors are common in call centres, when one call centre was developed, a holistic preventive strategy on stress at work was part of the development process. Preventive measures included the design of the work environment and work resources, and ensuring adequate job rotation and varied work (European Agency, 2002b).

Gender pay gap

Men earn more than women, even after figures are adjusted for number of hours worked. Although the degree of wage inequality has lowered among workers in the EU, women still remain over-represented in low-income and low/medium-income levels and they are seriously under-represented in the high-income categories. The gender pay gap is smaller in the candidate countries (Paoli, 2002).

Gender segregation in occupations

It is particularly important to look closely at where women and men work, as different hazards and different working conditions are found in different jobs and tasks, resulting in different health outcomes. A general overview of gender segregation in employment is given in this section. The next part of the report will then look in detail at gender differences in exposure to various hazards and health outcomes, and the issue of gender segregation will be relevant for discussing the gender differences to specific exposures and health outcomes.

Gender segregation refers to women being under-represented in some jobs and over-represented in others relative to their percentage share of total employment. Horizontal job segregation refers to the fact that women and men work in different employment sectors, and vertical segregation refers to the fact that even within an occupation women and men occupy different positions in the hierarchy. There can also be segregation within the same job, where women and men with the same job title perform different tasks. Statistics and surveys show that there is a high degree of gender segregation in the EU employment structure, even in countries where there are high employment rates of women.

Box 4

Horizontal gender segregation

Women are over-represented in:
- a limited range of occupations and industrial sectors;
- the public sector;
- small, private sector firms.

Vertical gender segregation

Women are under-represented in:
- higher status and higher paid jobs;
- senior management, senior government, and many professional occupations;
- higher graded jobs within one company.

Horizontal segregation

Below is a summary and illustration of areas of occupational segregation in the EU based on European Foundation (2002c) and Fagan and Burchell (2002) (see also Figure 7 and Tables 1, 2 and 3 in Annex 5):

- There is a high concentration of women in the public sector.
- When employed in the private sector, women tend to work in small and medium-sized companies.
- Female-dominated work areas include: services in private households, health, education and other care-related activities, sales, and hotels and catering.
- Male-dominated work areas include: construction, manufacturing, transport, agriculture and financial services.
- Only a small proportion of employment can be considered 'mixed'.
- Relative to their overall share of employment, women are disproportionately represented in caring, nurturing and service activities.
- Relative to their overall share of employment, men are disproportionately represented in management, and manual and technical jobs

associated with machinery or physical products considered to be 'heavy' or 'complex'.
- Men hold 80 % or more of jobs in the armed forces, the craft and related trades and plant and machine operator jobs.
- In manufacturing and craft trades, women's presence is low. Only in food and textiles, craftwork and as machine operators and assemblers do they have a presence of around 30 %.
- Men hold more than two thirds of skilled agricultural and fishery jobs.
- Two thirds of clerical and service and sales workers are women.
- Men predominate in engineering professions and women predominate in health and education professions.
- In the elementary professions, women are disproportionately represented in cleaning and agriculture-related jobs and men, in general, in labouring.
- Although there is country variation, the average picture of occupational segregation for EU-15 is largely replicated at the individual country level.
- Gender segregation is less prevalent in the candidate countries (additional data from a survey of 12 candidate countries — Paoli, 2002).

So, we find that typical employment areas for women include: clerks and secretaries, teachers, shop sales workers, cooks, catering assistants and waitresses, nurses and care assistants, textile machine operators, food processing jobs, fine assembly work, hairdressers and beauticians, cleaners and domestic workers. While women are employed less in the manufacturing sector, in the areas where they do predominate, such as textiles and food processing, it can be said that these areas of production and manufacturing are not considered as important as more male-dominated areas such as car manufacturing. In addition, within food processing, for example, women may work on the production line, while men may work in the stores or drive lift trucks.

Vertical segregation

Although, across the EU, women have increased their representation in management, men still dominate the top part of the occupational hierarchy. For example:
- 63 % of the workforce have a man as their immediate supervisor and 21 % have a woman;
- men hold more than 60 % of legislative and managerial occupations;
- more than 70 % of corporate managers and senior government officials are men;
- two thirds of the self-employed are men, and this proportion increases for the self-employed with employees.

Sources: European Foundation (2002c); Fagan and Burchell (2002).

Although vertical segregation remains high in the candidate countries, women are more likely to be in managerial positions in the candidate countries than in the EU (Paoli, 2002).

Vertical segregation means that, for example, although women predominate among nursing staff, men are over-represented in nurse manager positions. Likewise, in food processing and textile manufacturing, women predominate in the routine, production line jobs, but men are more likely to fill the supervisory, managerial and technical jobs.

Task segregation

Task segregation refers to women and men with the same job title carrying out different tasks: not only are women and men working in different sectors, and doing different jobs within the same sector or same workplace, but there is evidence that, even when they are employed in the same job within the same workplace, women and men may often carry out different tasks.

For example, researchers from Cinbiose, Quebec University, have found this to be widespread. One case was that of hospital cleaners where female cleaners carried out so-called 'light' tasks and men were assigned 'heavy' tasks. On closer inspection, the light work proved not to be so light. The women performed a wider variety of tasks and their postures were more variable. The workload was high, with postural constraints, repeated movements, a constant work pace and very little rest time. Having to work in static postures, and in bent or stretched positions was also frequent. For example, women had to stoop to empty many bins in a short time period, removing bags that could weigh more than 10 kg. In contrast, many tasks assigned to men, such as sweeping, were carried out in less tiring, upright positions. In another case involving gardeners, they found that women do more planting and weeding and men do more pruning. This could mean that they have different exposures to pesticides (Messing, 1998, 1999).

In other words, men may be assigned to tasks where the risks are more visible, whereas the women may be assigned to more repetitive tasks, possibly with more awkward positions (Messing et al., 1994, 1998a).

Conclusion

We have seen that the conditions of women's employment can vary from men's in various ways, including in relation to type of work, type of contract and domestic responsibilities. These differing conditions affect health and safety at work. Their influence will be considered further in the next section, which includes an examination of gender differences in exposure to various work hazards and health outcomes.

3.

GENDER DIFFERENCES IN OCCUPATIONAL SAFETY AND HEALTH EXPOSURES AND OUTCOMES

Introduction

European statistics and surveys suggest the following:
- Backache, stress, muscular pains in the neck and shoulders and overall fatigue are the most common work-related health problems in the EU.
- Men are more likely to have work-related accidents (of four days or more absence) than women.
- Women are more likely to report work-related ill-health problems than men.
- Women are more likely to report work-related upper limb disorders, work-related stress, infectious diseases, and skin problems.
- Men are more likely to report heart diseases, hearing problems and breathing and lung problems.
- Women are more exposed to intimidation in the workplace.
- Many working conditions are more closely related to occupational position in the occupational hierarchy or sector than gender per se.

Sources: Dupré (2002); Fagan and Burchell (2002); Paoli and Merllié (2001). See also Annexes 6 and 7.

This section of the report examines these gender differences in exposures to different work-related hazards and health outcomes, what they mean and why any differences seen may be occurring. Exposures and health outcomes in some particular sectors, such as agriculture, and for some particular groups, such as older women, are considered, as well as the situation outside Europe. The extent to which gender differences in exposure and health outcome relate to the gender differences in employment conditions, described in the previous section, are examined.

Accidents

Key points

- Men are more likely to suffer major or fatal accidents at work than women. Their serious injury rate is almost three times higher. When adjusted, for example for hours worked, it is still over twice as high. The gap between the accident rates of men and women is reduced further when less serious accidents are also included, but the rate for men is still over 1.33 times higher.
- The work-related accident rate is falling faster for men than women.

- Women and men suffer different accident rates and also different types of accidents. These differences are strongly related to job and task segregation.
- Women's accident rate is higher in fast-paced work and higher demand work — so tiredness and exhaustion may contribute.
- Using work equipment, machinery and tools designed for men may contribute to women's work accident risks.
- Accident statistics should take account of the number of hours worked. It would also be better to have a closer breakdown of statistics according to the real tasks people carry out. Practical means of improving occupational safety and health statistics by gender should be investigated.
- It would be helpful to have more information on types of accidents suffered by gender and a more detailed picture by Member State.
- More action is needed to prevent all accidents, both serious and minor, especially as simple control measures are often available.

Men are more likely than women to have serious accidents at work (for example, an accident resulting in a person being off work for four days or more). According to the figures for officially reported accidents in the Member States, compiled by Eurostat for 1998, men were almost three times more likely to be involved in accidents at work than women. These figures do not take account of number of hours worked. After adjusting for full-time equivalents in employment as well as standardising for differences in hours worked, the difference in accident rates between men and women decreases, but the ratio still remains higher than 2 (2.2) for the EU. These adjusted figures varied from over 2.5 times as likely for men to have an accident in Belgium, France, Austria and Portugal to only around 1.5 times as likely in the UK and Sweden (Dupré, 2002. See also Annex 7).

The 1999 EU labour force survey also included less serious accidents (resulting in under four days' or no absence). According to the responses, and after making the adjustments described above, the incidence rate of accidents at work was still over 1.33 times greater for men than women. On the other hand, according to the survey results, again after making the adjustments, women were 1.5 times as likely as men to suffer work-related health problems other than an accident in the EU in 1999 (Dupré, 2002. See also Annex 7).

In the third European survey on working conditions, 9 % of men and 6 % of women reported having been absent from work in the previous 12 months due to an industrial injury. In the same survey, women reported being less exposed to physical and material risks than men. However, the exposure is highest for men in manual jobs, followed by women in manual jobs. For non-manual jobs, the difference is negligible, although still slightly higher for men. In contrast to absence due to accidents, 10 % of both employed men and women had been absent due to health problems they attributed to their working conditions. Both male and female part-timers were less likely to be absent due to either work-related accidents or ill health, demonstrating the exposure effect (Fagan and Burchell, 2002. See also Table 7 in Annex 6).

The Eurostat adjusted figures confirm that both sector and type of work being carried out have an influence on the gender difference. Men are more likely to be employed in the high-risk sectors of construction and transport, whereas the accident risk is lower in offices and shops, where women are disproportionately employed. However, looking at the rates between men and women within sectors, the difference is much greater in construction and in the energy and water industries, where men and women are strongly segregated into different work areas — for example, in construction, where men are working on the sites and women in the offices. However, in the hotel and restaurant sector, where it can be presumed that women and men are carrying out much more similar tasks, men only have a marginally higher accident rate. In the finance and business sector, the accident rate was over 1.5 times higher for men, but male-dominated activities such as security staff are included in this sector.

For example, in Sweden, among women, the highest risks of serious occupational accidents are in the sawmill industry, public transport, regular bus traffic and the plastic packaging industry. All these industries have between two and three times the average work-related accident risk (Broberg, 2001).

Box 5: Causes of injuries in the restaurant and catering sector

More women than men are employed in sales, hotels and catering. In hotels and restaurants, the accident rate is similar for men and women. An analysis of the causes of accidents reported in the catering sector, including hotels and restaurants, in Great Britain found the following, by frequency of occurrence:

Slips and trips	30.0 % of reported injuries (75.0 % of major injuries)
Handling	29.0 %
Exposure to hazardous substances, hot surfaces, steam	16.0 %
Struck by moving/falling objects, including hand tools	10.0 %
Walking into (mainly fixed) objects	4.0 %
Machines, e.g. used for slicing and mixing food	3.0 %
Falls	1.8 %
Fire and explosion	1.6 %
Electric shock	0.5 %
Transport, e.g. lift trucks	0.3 %

Source: Health and Safety Executive (1997), *Priorities for health and safety in catering activities.*

Slips, trips and falls in the education sector

The table below, from Great Britain, shows that slips and trips are also a major cause of accidents in the education sector. Most injuries result in strains and sprains, but they also include broken bones and head injuries. Some accidents can even result in fatalities as a result of complications. They have not always been taken seriously, partly because they are thought to be inevitable, although there are simple prevention measures.

Injuries due to slips and trips in the education sector

Reported injuries (to British OSH authority 2001/02 (provisional)	Total reported injuries	Employee injuries due to slips and trips	% of total injuries due to slips and trips
Primary and secondary education	3 700	1 399	38
Higher and further education	1 793	643	36

Ways of controlling slip and trip risks include the following:

- Preventing slip and trip risks at the design stage by ensuring appropriate flooring with a slip-resistant surface.
- Cleaning up spillages immediately and ensuring floors are dried after cleaning.
- Establishing a sensible shoe policy for all staff and students.
- Providing adequate lighting. Poor lighting or glare can obscure slip and trip hazards.
- Information and training — control measures that involve changes to working practices can be as equally effective as physical control measures.

Sources: Health and Safety Executive (2003a, 2003b)

Job segregation also has an effect on the types of and reasons for work-related accidents. For example, a Swedish study on slips and falls concluded that women may seldom fall from scaffolding but it is typical of them to slip when trying to prevent a patient, pupil or customer from falling (Kemmlert and Lundholm, 2001). Grönkvist and Lagerlöf (1999) quote an example from the food sector where female employees more frequently suffered injuries from machinery, while men more frequently suffered injuries from sharp tools, such as knives, and contact with animals, reflecting their different tasks.

It would be useful to have from the Member States not only information on the relative numbers of accidents by gender, but also on the different types of accidents and covering both fatalities and over four-day absence accidents as well as less serious accidents. Where this information can be made available also by sector or type of job, it is even more useful.

According to data from the third European survey on working conditions, women have a higher accident rate where the pace of work is intense or the work is more demanding, involving working to tight deadlines (Paoli and Merllié, 2001). So stress, tiredness and exhaustion may contribute. Men, however, are more likely than women to have their pace of work set by products and machines. Women are more likely to have the pace set by customers and service users (Fagan and Burchell, 2002).

Women are slightly more exposed than men to the insecurity of temporary contracts. There is evidence that workers with non-permanent contracts have more accidents than those employed on a permanent basis (Notkola and Vänskä, 1999; see also Goudswaard and Andries, 2002). The increased risk could be due to the differing nature of work, less training and less knowledge of the particular workplace or tasks. However, men are more likely to be self-employed and the work-related accident rate is higher for self-employed persons.

According to Eurostat figures, the overall EU work-related accident rate is falling and the reduction is greater for men than women. Several reasons are suggested for this: a greater shift for men away from high-risk, traditional manual jobs; greater improvements in safety in the workplaces where men work; or safer working practices adopted by men themselves, as their jobs and workplaces are more likely to be the target of efforts to reduce accidents through campaigns, etc. (Dupré, 2002. See also Annex 7). It appears that further efforts are needed to prevent accident risks to men, but that attention should also be paid to accident risks to women and there are examples of campaigns to tackle accidents in sectors where women are more prevalent (see Box 6).

Box 6

Recipe for safety

A national OSH authority campaign in the food and drink sector set key national objectives to reduce accidents. This was backed up by guidance, involvement of trade associations and trade unions, an agreed strategy for labour inspections, and targeting company sites with the highest accident rates for special support in improving health and safety management. Issues targeted included manual handling, slips and trips, and machinery accidents.

Preventing needle injuries to sewing machine operators

A clothing company worked closely with a trade union to find a way to protect operators' fingers from being punctured by the machine needle during sewing operations. The solution was to design a needle guard. The guard has subsequently been the basis for the development of a new CEN sewing machine safety standard.

Source: European Agency (2001a).

The use of work equipment, machinery and tools designed for men may contribute to women's work accident rate. For example, Messing (1998) cites an example of pruning trees, where women could not be assigned the task because 'the belt of the device that hoisted pruners into the air was too large for most women, and they could slip through'. One study area that has looked at comparative injury rates between women and men where they are required to carry out tasks under the same conditions has been training in the British Army. These identical training requirements were brought in to comply with equal opportunities legislation. According to military medical statistical reports, women in the army are at increased risk of injury and musculoskeletal disorders (Bergman and Miller, 2001; Strowbridge, 2002; see also Geary et al., 2002). Presumably, the training methods and equipment used were designed for the male norm. Strowbridge (2002) concludes 'these results may act as a basis for targeted intervention in order to reduce inequality without reducing overall training standards'. Strowbridge also mentions evidence that female recruits may be more likely to report injuries than male colleagues.

Men are also involved in accidents through poor fit between work requirements and the person. If men are assigned more heavy lifting than women because they are considered fitter, they will consequently suffer more lifting accidents. Therefore, the best approach is to eliminate risky processes where possible and make the job accessible and safe for more people.

The main conclusion is that men have more work-related accidents whereas women suffer more work-related ill health. In research as well as in policy, labour inspectorate activities and other actions, priority is often given to 'high-risk' accident areas, which will mean that more health and safety resources will go into areas that affect men than areas that affect women.

At the very least, when setting priorities, overall numbers that would benefit from action should be taken into account as well as accident rate; for example, larger numbers employed in the hotel and restaurant sector or the very large numbers employed in offices compared to the sea fishing sector. This is not to say that it is not right to target the high-risk sea fishing sector, but, in addition, improvement in a lower accident risk, high-employment area would benefit the safety and health of a very large number of people.

As described above, there are gaps in information. It is also clear that, although in each Member State men have more accidents than women, the picture of accident rate is not uniform. For example, according to the figures from Eurostat, the rate of serious accidents among women rose in Spain between 1994 and 1998, perhaps because more women now have jobs in a high-risk sector, or that safety standards may be lower (Grönkvist and Lagerlöf, 1999). The rate also rose in Belgium, Luxembourg and Sweden, although it fell in other Member States. More information would also be useful on differences in the comparative male/female accident ratios between Member States, including differences by gender in types of accidents or jobs where accidents are occurring. It would be useful if all Member States carried out more detailed gender analysis of their accident and ill-health data, gathered supplementary information through labour force surveys and carried out individual studies where gaps in information are found.

Improved information about the types of accidents suffered by gender, both in the serious and less serious categories, would be useful. It would be better to have a closer breakdown of statistics according to hours worked and according to the real tasks people carry out. In conclusion, practical means of improving occupational safety and health statistics by gender should be investigated.

Box 7: Is there a gender difference in accident rates?

According to Eurostat figures and the European survey on working conditions, males have more accidents than female workers, even when adjusted to take account of differences in hours worked. In Great Britain, to supplement official occupational accident reports, some health and safety questions are asked in the national labour force survey. The Health and Safety Executive performed further analysis of the work-related accident data from the survey on various issues, including gender (Health and Safety Executive, 2000).

The overall rate of workplace injury according to the survey data is over 75 % higher for men than for women. However, after allowing for the effects of occupation, hours of work and other characteristics, this rate drops considerably to a 20 % difference. Job characteristics explain much of the difference, but men still have a 20 % higher relative risk not obviously explained by job characteristics. Regarding age, the Health and Safety Executive reported that men aged 16–20 have a substantially higher risk of all workplace injury than older men. For women, there is no significant variation between the age groups. However, the sample size for women reporting injuries was too small to see if there were any statistically significant differences in risk by age group.

Are women safer workers than men? It is unlikely to be as simple as that. Another question that has to be asked is 'Are the men and women employed in the same jobs really doing the same tasks?'. Messing (1998) has found that women and men, even those in the same jobs, are not necessarily exposed to the same health risks. She concludes that there are not only biological differences between working women and men but they also have different levels of job seniority, average age, work assignments, work techniques and lives outside work.

Musculoskeletal disorders

Key points

- In general, in Europe, musculoskeletal disorders (MSDs) are a serious and growing problem. They are the most reported work-related health complaint among both women and men in Europe.
- Women and men report similar levels of musculoskeletal complaints. However, women report more upper limb pain.
- Musculoskeletal disorders occur in relation to exposure to poor ergonomic conditions in both women and men. Women often work in conditions associated with musculoskeletal disorders — work requiring awkward postures, monotonous and repetitive tasks, inappropriate work methods and work organisation and, more often than is commonly recognised, heavy lifting.
- There is adequate scientific knowledge regarding specific occupational ergonomic risks to prevent a large proportion of musculoskeletal disorders among working people. Preventive measures should not and need not be discriminatory with respect to gender.
- Despite the widespread belief that musculoskeletal disorders are more common among women, the issue has been studied more among men. Further study of prognostic factors among women is needed as well as more studies to investigate whether risk varies between women and men with the same occupational exposures.
- Some tiring aspects of work that are more a feature of women's work, for example standing, have been studied much less than, for example, lifting weights.
- It could be helpful to distinguish between types of MSD in reporting systems, especially back injuries from manual handling and upper limb disorders.

Musculoskeletal disorders cover a broad range of health problems. The main work-related groups are back pain and injuries and work-related upper limb disorders which include the neck, shoulders, elbows, arms, wrists and hands. Lower limbs can also be affected, and 'housemaid's knee', a condition associated with kneeling work, for example to scrub floors, was well documented in the past. Limb disorders include a wide range of inflammatory and degenerative diseases and disorders that can affect tendons, ligaments, nerves, muscles, circulation and joint cartilage and result in pain and functional impairment. Back disorders include spinal disc problems and muscle and soft tissue injuries. For most musculoskeletal disorders, both onset and duration are more often chronic or sub-chronic than acute. Typically, they tend to develop after months or years of overuse of the soft tissues (Punnett and Herbert, 2000).

European surveys show that musculoskeletal disorders are the most common work-related health problem for both men and women and that they are on the increase. Contributory factors such as intensity of work have also increased. Around 30 % of women and men report muscle-related problems from their work (back and other muscular pains). If the figures are compared for full-time and part-time work, women report more work-related problems in each group, although the difference is negligible (Dupré, 2001; Fagan and Burchell, 2002). As regards only neck, shoulder and upper limb pains, these were reported by a quarter of respondents to the third European survey on working conditions. There was no significant difference in response between women and men, except that women reported significantly more upper limb pains. Women and men reported similar amounts of work-related back pain in the same survey (Paoli and Merillé, 2001; see also Punnett and Herbert (2000) regarding increased incidence of upper limb disorders among women workers).

Work-related upper limb disorders are associated with poor postures, highly repetitive movements, forceful movements, fast-paced work, hand–arm vibration, work in cold environments and psychosocial factors including work-related stress. Back disorders are associated with manual handling work, where risk factors include the weight of the load, frequency of manual handling work and awkward postures. They are also associated with awkward postures in general, including prolonged sitting and poor seating, as well as prolonged standing. These risk factors are found in many of the jobs that women carry out. Exposure to whole-body vibration, particularly when seated, is another factor linked to back pain and is more common in male-dominated jobs such as driving trucks and buses. In the third European survey on working conditions, women and men reported similar exposures to repetitive work and postural loads, and working in painful and tiring positions and men reported more exposure to heavy lifting. In the previous survey, women reported more exposure to repetitive work than men, and it is suggested that there may have been some confusion over the interpretation of the question in the third survey (Paoli and Merllié, 2001).

Comparisons between work and gender factors regarding the incidence of work-related upper limb disorders frequently find strong association with workplace risk factors (Buckle and Devereux, 1999). Punnett and Herbert (2000) are among those who have reviewed incidences of musculoskeletal disorders among women workers: for example, high prevalences among women, linked to work-related factors, have been found among garment workers; women workers in packing jobs; textile manufacturing; electronics manufacturing; assembly line manufacturing; poultry and fish processing; VDU and telephone operators; cleaners; kitchen workers; and cashiers in supermarkets. Female-dominated packing and assembly jobs, for example, typically require rapid, precise, repetitive movements and prolonged sitting. Even when men are also employed in this type of work, with the same job title as women workers, their jobs may still differ, with the women being more exposed to such risk factors (see also Messing, 2000). Studies that find job demands predictive of neck and shoulder pain include Ariëns et al. (2001).

> **Box 8: Musculoskeletal disorders among textile and garment workers**
>
> Repetitive work and other MSD risk factors are prevalent in the clothing industry, including pay systems traditionally linked to work pace. The industry has been forced into wholesale restructuring in recent decades by fierce global market competition. The resulting intensification of work has led to worse working conditions and a higher incidence of MSDs. However, participatory approaches have been successfully used to address established and emerging MSD problems in different workplaces across Europe:
>
> - A research intervention in a ladies' underwear manufacturer and packer in Italy resulted in: a guide to prevention, sponsored by the company and other agencies; recognition of the need to continually monitor for MSD problems, due to changes taking place within the business; a proposal to develop simple tools for workers themselves to use to evaluate working practices.
> - A jeans manufacturer in Spain tried to introduce a 'modular manufacturing system'. However, in practice, operators still appeared to be working on single tasks and had not received training in team working. The jeans marketplace is very difficult and the enterprise was also hoping to improve its competitiveness with the changes, but more attention needed to be paid to ergonomic factors and broader work organisation issues. One barrier to success was a lack of access to innovative solutions and sharing of experiences.
> - Danish trade unions and employers developed a national action plan to support prevention of MSDs in the sector. The plan was provided with a budget. A manual was produced on alternative wage systems to piecework. Goals were set to change work organisation and introduce team working and pilot studies were carried out.
>
> *Sources:* Hague et al. (2001). See also Barros Duarte et al. (2002); IDICT web site; European Agency (2001a).

While, according to the third European survey on working conditions, men do more manual handling work, the number of women engaged in these tasks is not insignificant: 28 % of men report spending at least half their time at work carrying or moving heavy loads compared to 17 % of women.

Back disorders from heavy physical work have been documented among nursing and care workers, cleaners, kitchen staff and laundry workers. It is often not appreciated that in some jobs women can carry out significant heavy, physical work. For example, the National Institute of Occupational Safety and Health in the

> **Box 9: Homecare health workers — a growing and mainly female service with a growing rate of musculoskeletal problems**
>
> NIOSH was concerned with the growing rate of musculoskeletal injuries, particularly back injuries, among healthcare workers in general and workers delivering care services in patients' own homes in particular. While in the United States the rate of overexertion injuries among healthcare workers is around double the national average, in the case of home healthcare workers it is even higher — around three times the national rate. In addition, it found that very little research had been conducted to address health issues for home healthcare workers, although in the United States this is the fastest growing part of the healthcare industry. Compared to the hospital setting, the NIOSH researchers found that some risks factors were magnified in the home healthcare setting such as: working alone; no control over arrangements of furniture/equipment; non-adjustable beds; lack of patient transfer equipment; and patient incapacity levels increasing. Patient care policies and privacy issues add further complications. Reviewing the research literature and existing intervention studies, they have concluded that: the high prevalence of musculoskeletal injuries is due to poor ergonomic working conditions, which result in workers having to use awkward postures and forceful exertions during patient care tasks; ergonomics — the design of all components of the work environment to better accommodate the capabilities of the worker — is the most promising prevention approach; ergonomic solutions specific to the home setting are needed; successfully implemented ergonomics programmes in this setting need to be documented and shared. On the basis of this research, NIOSH will conduct further research to quantify the risks to home healthcare workers and intervention work aimed at prevention of musculoskeletal disorders and other risks (see Galinsky et al., 2001).

United States (NIOSH) reports that the rate of overexertion injuries among healthcare workers, particularly associated with patient handling, is over double the national average, while for homecare workers it is around three times the national rate — see Box 9. A study by Dassen et al. (1990) found that female nurses performed more heavy lifting of patients compared to male nurses, who were more likely to be engaged in managerial tasks or in surgery or intensive care areas where the lifting work is less. Female shop staff and agricultural workers may have to lift significant loads and female agricultural workers may also carry out heavy, physical work and work in awkward postures, such as stooping positions. Even in offices, women may lift heavy folders of records, boxes of photocopy paper or office equipment. In addition, when the jobs of some women are examined more closely, their 'light work' is often more physically demanding than it initially appears, not only due to weight, but also due to the repetitiveness and awkwardness of tasks and poor postures that have to be maintained to carry out the work (for example, Messing, 1998 and 1999, gives an example of the workload of cleaners emptying bins).

Research evidence suggests that using a computer for six hours or more per day appears to be associated with increased risk of work-related upper limb disorders. There is some evidence that this association is stronger for women than for men and symptoms can already be observed after four hours' computer use per day (Blatter and Bongers, 2002). Control over the pace of work and ability to take regular rest breaks are important factors in prevention, along with suitable workstations and layout.

It is increasingly suggested that psychosocial factors such as intensive work, low job control, low social support, monotonous work and other factors that contribute to stress at work play a part in the development of work-related upper limb disorders (for example, see Devereux, 2000; Vingård and Hagberg, 2001). As is discussed in the section on work-related stress, these features are common in the work of many women.

Musculoskeletal disorders and the design of work equipment

Musculoskeletal disorders can affect workers in all sectors but an additional risk factor for many women is the use of tools and equipment not always designed for a female work population (Punnett and Herbert, 2000). The two studies presented in Box 10 demonstrate this. One concerns musculoskeletal disorders among women in a non-traditional work area, gardening, where one might expect to find that equipment has not been designed for a female population. The other concerns the work of a predominantly female population, cleaners, but here unsuitable work equipment was still found. Paying more attention to the ergonomic fit of work tools and tasks to the worker's physique should not only benefit women in non-traditional work areas, but also men who are not of average build (Messing, 1999).

Box 10: Municipal government gardeners — an example of women in 'non-traditional' jobs using poorly adapted tools

Many tasks were found to cause physical difficulties for both men and women. Problems found in a study for both male and female municipal government gardeners included the following:

- Problems where women and some men differed from the male norm used to design work tools and equipment.
- Problems related to physical build. Women reported similar problems to smaller men about jobs requiring strength like handling loads and digging.

- Women and tall men often reported problems with equipment and tool handling, and postures and positions.
- Problems particularly for women with protective equipment, tools and appliances and vehicle handling, due to design factors (too big, tractor seat too high, etc.).
- Differences in pains experienced by men and women can be explained partly because of different tasks performed (although employed in the same jobs), different working methods and differences in body build.

The proposed solutions include the following:

- Tackle problems which are too strenuous for all workers, for example by use of team work and changes to work methods and work equipment.
- Supply a wide range of tools, or adjustable tools.

Source: Summary of case study in Messing (1999).

Musculoskeletal disorders and the design of work equipment used by cleaners

A UK study looked at musculoskeletal problems in cleaners, a predominantly female workforce and often an older workforce (Woods et al., 1999). Work factors were found to be associated with an increased risk of musculoskeletal disorders among the study group. These included unsuitable design of equipment, specific work tasks and postures and poor work organisation and lack of training. Recommendations were made on machines and equipment redesign, changes to work organisation and work schedules, and development of local health-monitoring policies and equipment maintenance programmes. The study looked at the use of floor buffing or polishing machines, mops and vacuums. The main areas of concern regarding the equipment were: lifting and moving machines; unsuitable handle shape, size and angle and difficulty with handle adjustment; forces required to operate the machines and use equipment; vibration from machines; inadequate machine maintenance.

Recommendations included some aimed specifically at designers and manufacturers: buffer-machine weight, height control, controls, vibration and discs; mops — length, bucket stability, manual handling, squeezing system; vacuum cleaners — weight, handle length and design, location of controls.

The researchers concluded that the design of equipment should accommodate the physical dimensions and strength of a wide range of potential users including the smaller and weaker members of this workforce. In particular, the design should take account of the problems that affect cleaning personnel (and that are substantiated by objective measures). The results of this research were followed up by a number of initiatives including: a seminar to discuss the findings with employers, trade unions, equipment manufacturers and the national labour inspectorate body; and a project to obtain case studies of good practice. The research has been used to produce national guidelines on prevention.

Sources: Woods et al. (1999); Health and Safety Executive (2003c).

Musculoskeletal disorders and gender — gaps in information

There is a lack of comparative information on musculoskeletal disorders, as they cover a range of conditions and not all incidents are officially recognised as industrial injuries, nor is there consistency across the Member States. This affects what data are collected and the apparent incidence of different types.

It has already been mentioned that even where men and women have the same job title, they may carry out different tasks (Messing, 1999) and therefore will have different exposures to occupational musculoskeletal risks. Many studies fail to take account of this.

Standing is common in many jobs dominated by women, such as shop assistants, teachers and hairdressers. However, there are far more studies on lifting in relation to work than on standing (Messing, 1998). Standing involves static muscle effort. Maintaining any position for a period of time is tiring and a known contributing risk factor for musculoskeletal disorders.

> **Box 11: Prevention of musculoskeletal disorders among hairdressers**
>
> Hairdressing is a typical female-dominated work area and musculoskeletal problems are a reason for absenteeism. Risk factors include repetition and exposure time to tasks (daily and weekly working time and frequency and duration of breaks) as working conditions are often very hectic with limited opportunity for rest breaks; design and handling of work tools and equipment; work posture, including excessive reaching, holding the arms in a raised position, and one-sided strains of the musculoskeletal system; standing; and the physical and psychosocial work environment.
>
> In response to the high rate of musculoskeletal problems — nearly 10 % of notifications of illness are caused by backache — the German health insurance fund AOK developed a cross-company approach, which included training provided through the vocational school in Bavaria. It has reduced absenteeism in the companies and reduced the need for AOK staff to visit them, and trainees learn about health promotion before adverse health effects appear. Since 1997, training workshops have been held annually within the vocational schools covering the following elements:
>
> 1. Theory (basics about the back and musculoskeletal disorders, occupational causes and health).
> 2. Practice (analysis of complaints at the workplace and exercises).
> 3. Review and refresher training (repetition of the most important elements and provision of individual advice in the last year of vocational school).
>
> In addition, a guide 'Healthy backs in the hairdressing business' was developed and is used for providing information to hairdressers.
>
> *Source:* AOK Bayern (2002).

There are also many unsolved questions concerning musculoskeletal disorders. Age, gender, socioeconomic status, race or ethnicity, history of acute trauma, various systemic diseases, smoking, alcohol use, obesity, and factors affecting female hormone levels, such as use of oral contraceptives, have all been linked to musculoskeletal disorders. But evidence concerning whether and how these factors may affect the occupational outcomes of musculoskeletal risk exposure is far from clear (Punnett and Herbert, 2000). Many research results are conflicting or the research methodology is problematic. For example, although many studies suggest that men have more musculoskeletal disorders, large cross-sectional sample studies may mask gender differences, as they often place more emphasis on whole-body exertions and energy expenditure than on localised repetitive stresses to the upper extremities typical of women's jobs (Punnett and Herbert, 2000) and jobs preferentially assigned to women may have specific exposures that escape the attention of researchers (Messing, 2000). It has also been suggested that women report more symptoms simply because they are better at recognising, articulating and communicating their musculoskeletal disorder symptoms than men. There is also some evidence that women consult a doctor quicker than men when health problems arise.

Conclusions

Despite some outstanding questions, there is now a significant body of evidence concerning the work-related causes of musculoskeletal disorders and their prevention. Based on a review of over 200 studies of musculoskeletal disorders, Punnett and Herbert (2000) have drawn a number of conclusions about what is known about gender and musculoskeletal disorders and what should be investigated further — see Box 12. Overall, we know that the occurrence of musculoskeletal disorders is strongly related to work-related factors in both women and men and there is adequate scientific knowledge to take preventive measures, based on ergonomic interventions, and that interventions do not need to be discriminatory with respect to gender. Analysis of gender as a 'risk factor' does not help the understanding of these issues. However, further research is needed in a number of areas, such as whether there are gender variations in musculoskeletal risk where occupational exposures are the same, and whether there are gender variations in exposure–response relationships.

> **Box 12**
>
> Punnett and Herbert (2000) carried out an extensive review of work-related musculoskeletal disorder research and gender. Their conclusions were as follows:
>
> - Musculoskeletal disorders occur in relation to ergonomic exposure in both men and women. There is substantial clinical, biomechanical, and epidemiological evidence supporting the relationship between musculoskeletal disorders and ergonomic factors in the workplace, including high repetition, high manual forces and vibration.
>
> - Thus, there is adequate scientific knowledge regarding specific occupational ergonomic stressors to prevent a large proportion of musculoskeletal disorders among working people. Preventive measures should not and need not be discriminatory with respect to gender. The best approach to eliminating musculoskeletal disorders from the workplace is the implementation of engineering controls, such as changes in equipment and workstation and job design, in the context of a comprehensive ergonomic programme with participation from all levels of the enterprise.
>
> - More research is needed to elucidate whether musculoskeletal disorder risk varies between women and men in jobs with the same occupational exposures, and whether work-related musculoskeletal disorders have the same outcomes in women and men. Women report musculoskeletal disorders more frequently than men; however, this difference appears to be less marked for lower back disorders and when men and women are compared within homogeneous job groups. Exposure–response relationships have been examined by gender in only a few studies, and the results have not been consistent. However, some studies suggest that men may have a higher risk than women with increasing exposure to physical stressors, although women have a higher background risk. This may mean that other factors have a greater effect on women in low-exposure jobs and are less important when there is high physical loading, or because women with higher occupational exposures (physical and/or psychosocial) are more likely than men to leave employment or change jobs due to work-related musculoskeletal disorders. In contrast, the sparse literature suggests that women may both experience different levels of job strain and have increased vulnerability to similar levels of job strain compared with men, possibly because of the added demands of household responsibilities. All of these explanations remain tentative at present and require further study.
>
> - Despite the widespread belief that musculoskeletal disorders disproportionately affect women, the outcomes of these conditions have been examined primarily in men. In order to tailor appropriate interventions for primary or secondary prevention of disability, further study of the prognostic factors among women is an important research priority.
>
> - Analysis of gender as a 'risk factor' for musculoskeletal disorders, or adjusting for gender differences, does not elucidate these issues. Future epidemiological studies of musculoskeletal disorders should include participants of both genders to avoid unnecessary constraints on the available exposure contrasts. The associations of musculoskeletal disorders with gender and occupational ergonomic exposures should be assessed separately in order to determine whether women are at increased risk when exposed to the same ergonomic stressors as men. Gender-stratified presentation of data is valuable because it permits examination, rather than smoothing over, of differences in the exposure–response relationships.

Work-related stress

Key points

- Stress is a major work-related health problem in Europe for both women and men.
- Women report more work-related stress health complaints than men.
- There are known causes of work-related stress and these factors are present in many jobs typically done by women.

- Women are more exposed to some specific stressors because of: the type of work they do; their position in the hierarchy of organisations; discrimination; sexual harassment; their situation outside of work.
- The established safety and health risk-assessment and management approach can be applied to preventing the causes of work-related stress.
- Both instruments used for research and risk assessment should cover work issues that affect women more, such as working conditions incompatible with family responsibilities, sexual harassment and discrimination.
- Work, as opposed to not working, is positive for the health of women.
- Further research to place stress at work in the context of other issues, such as gender, is essential in order to achieve a complete perspective for the management of stress at work.

What causes work-related stress?

Stress can be said to be experienced 'when the demands of the work environment exceed the employees' ability to cope with (or control) them. Defining stress in this way focuses attention on the work-related causes and the control measures required' (Cox et al., 2000). The resultant stress can take the form of emotional, cognitive, behavioural and physiological reactions and reactions to the same psychological exposures may vary between individuals.

Box 13: Stressful characteristics of work

Category	Sources of stress
CONTEXT OF WORK	
Organisational culture and function	Poor communication, low levels of support for problem-solving and personal development, lack of definition of organisational objectives.
Role in organisation	Role ambiguity and role conflict, responsibility for people.
Career development	Career stagnation and uncertainty, under- or over-promotion, poor pay, job insecurity, low social value of work.
Decision latitude/control	Low participation in decision-making, lack of control over work (control, particularly in the form of participation, is also a context and wider organisational issue).
Interpersonal relationships at work	Social or physical isolation, poor relationships with superiors, interpersonal conflict, lack of social support.
Home–work interface	Conflicting demands of work and home, low support at home, dual career problems.
CONTENT OF WORK	
Work environment and work equipment	Problems regarding the reliability, availability, suitability and maintenance or repair of both equipment and facilities.
Task design	Lack of variety or short work cycles, fragmented or meaningless work, underuse of skills, high level of uncertainty.
Workload/work pace	Work overload or underload, lack of control over pacing, high levels of time pressure.
Work schedule	Shift working, inflexible work schedules, unpredictable hours, long or unsocial hours.

Source: Cox et al. (2000).

While individuals may experience stress differently and manifest it in different ways, there are a number of factors that have been established as known sources of stress. These are summarised in Box 13 and there is a reasonable consensus based on both theoretical and empirical evidence that these factors increase the risk of work-related stress (Cox el al., 2000). In addition, studies have not found gender differences in sources of work stress (for example, Miller et al., 2000). Some specific sources of stress, exposure to violence from members of the public, and bullying and sexual harassment from work colleagues are discussed in subsequent sections.

Some health implications of stress

Stress can be linked to a number of mental and physical ill-health effects and experiencing continuous stress is harmful. Stress can increase the risk of heart disease and depression and it can also weaken the immune system, making us more vulnerable to illness. It can even lead to suicide. There are also links between stressful work and musculoskeletal problems (Ariëns et al., 2001; Bongers et al., 1993; Buckle and Devereux, 1999; Hoogendoorn et al., 2000; Punnett and Herbert, 2000). The many symptoms of stress include raised blood pressure, depressed mood, irritability, chest pains, digestive disorders, sleep disorders, and increased drinking or smoking.

Do women suffer more from work-related stress than men?

Stress is the second most frequently reported work-related health problem across Europe according to the third European survey on working conditions, with 28 % of respondents reporting that stress from work was affecting their health (Paoli and Merllié, 2001). Stress at work accounts for more than a quarter of absences from work of two weeks or more through work-related health problems in the EU, according to the European labour force survey (Dupré, 2001).

According to a gender analysis of the responses to questions about work-related illness and complaints in the European labour force survey, 'there was slightly less variation across the Member States among men reporting stress-related complaints, and, in each case, the proportion was less than for women […]. In the EU as a whole, among the victims of work-related health problems, 17 % of men and 20 % of women reported stress, depression and anxiety as being the most serious complaints. In the EU as a whole, stress-related complaints were the second most frequently reported complaint by both women and men' (Dupré, 2002). While overall, according to the third European survey on working conditions, there is little difference between men's and women's perceptions of stress-related problems caused by their work, once again part-timers reported fewer stress symptoms. According to the survey, the highest levels of stress were experienced among professionals, technicians, managers, machine operators, service workers, clerks, and craft workers.

The sources of stress related to work given in Box 13 can be present in the jobs of both women and men. However, women may be disproportionately exposed to these factors because of job segregation and their increased caring and home responsibilities. In addition, sexual harassment, discrimination, blocked career advancement, and disrespectful treatment are stressors more common in women's working lives than in men's working lives (Landrine and Klonoff, 1997). Additional stressors for women in professional or managerial jobs include organisational politics, overload, roles and expectations concerning social-sexual behaviour, and reconciliation of work and home (Nelson and Burke, 2000). Lack of control over one's own work is well recognised as a work-related stressor (for example, Karasek and Theorell, 1990). Results of the third European survey on working conditions suggest that women may be more exposed to this stressor than men. This is consistent with the results of research studies, for example Vinke et al. (1999).

> **Box 14**
>
> Responses to the third European survey on working conditions suggest that, compared to men, women are:
>
> - less likely to have planning responsibilities in their jobs;
> - more exposed to monotonous tasks;
> - less likely to work in jobs involving problem solving and learning;
> - less able to choose when to take breaks;
> - more likely to have their work interrupted to deal with unforeseen tasks;
> - less likely to receive training.
>
> What is more, women in professional jobs have lower work autonomy than male professionals.
>
> *Sources:* Paoli and Merllié (2001); Fagan and Burchell (2002).

Many studies find that employed women have higher levels of stress and distress than employed men (Williams and Umberson, 2000). Responsibility to multiple supervisors, for example in clerical work, and unclear work expectations are stressors common in female-dominated industries, especially in the service sector. Also, overtime work and unpredictable or inflexible scheduling are stressors that women with caring responsibilities suffer particularly. Monotonous work and low control are characteristics typical of many female-dominated occupations. Miller et al. (2000) found little gender difference in sources of workstress among managers, but found that female managers experienced more distress. They hypothesise that the greater psychological and physical ill health reported by women is due to work/home overload and conflict. In a study of female and male managers, Davidson et al. (1995) found that female managers were under much more pressure than their male counterparts, for example stemming from 'organisational structure and climate' and discrimination and prejudice.

Employment in emotionally demanding work, or 'people work', is more common among women than among men. Examples of this emotional labour include nurses working with dying patients or dealing with distressed patients or relatives and teachers working with children with learning difficulties. Workers in such jobs are prone to stress-related diseases, depression and burnout (for example, Houtman and Dhondt, 1994). Shift working is an additional stressor in this type of work that, apart from disruptions to body rhythms, can add to work–life conflicts.

Ballard et al. (2002a, 2002b) studied commercial airline flight staff, commenting that, although this group had been the subject of many studies investigating cancer, little attention had been paid to psychosocial risks. They found an elevated rate of death by suicide among female flight attendants. Women cabin crew reported: concerns about the adequacy of their role as mothers and in the family as well as maintaining social relations; limited time to attend to problems in their personal lives; and feelings of loneliness and isolation at work and outside. Many reported panic attacks, anxiety and depression, experienced either personally or by colleagues.

In relation to the occurrence of work-related stress, we come back once again to job segregation, task differences and differing situations. The typical jobs of many women mean that they do more repetitive work and have less control over their work than men. As already mentioned,

> **Box 15**
>
> 'Disruptive interruptions [to work tasks] are associated with higher levels of work-related illness for both men and women, but the effect on women is more severe than the effect on men. The same is seen for intensity of work, where again the healthiest group are the women who work in jobs with the least time pressures, and the unhealthiest group are the women with the most time pressures at work. The same pattern was also evident for working unsociable hours and variable hours. The evidence [...] suggests that women have more to gain from an improvement in working conditions.'
>
> 'When differences in men's and women's working conditions and occupational position are controlled in the analyses [of the third European survey on working conditions], we found that women were more susceptible to work-related ill health than men. This may be partly due to domestic workloads that many women carry out. It may also be because there are other working conditions that women are disproportionately exposed to that are not picked up by the [...] survey. This issue requires further analysis and consideration in light of the current review of the EU regulatory framework on health and safety.'
>
> Excerpts from the analysis by gender of the third European survey on working conditions, Fagan and Burchell (2002).

many female-dominated jobs are emotionally demanding (Hochschild, 1983; Boyd, 2002). In addition, nursing work often involves shift work, which is known to cause social and psychological problems (Hatch et al., 1999). Also, women's double workload may increase the risk of stress-related psychological disorders such as chronic fatigue, nervousness, anxiety, sexual problems and depression (Wedderburn, 2000).

Some differences in reported reasons for stress are related to vertical job segregation. For example, in a study of managerial, professional and clerical university employees, Vagg et al. (2002) found that organisational-level effects were both more numerous and larger in magnitude than gender effects. Employees at higher levels reported that they experience stress more often while making critical decisions and dealing with crisis situations than did workers at lower levels, for whom inadequate salary and lack of opportunity for advancement were more stressful. There were some gender differences found: for men, work stress was more strongly

> **Box 16**
>
> Research carried out at an Italian suit (garment) manufacturer found a connection between stress, MSD and experience of pain. The higher the stress levels reported by workers, the greater the pain experienced from their work-related injuries (Hague et al., 2001).
>
> An Italian study reports data collection on socio-demographic characteristics, relationship between mental health and working environment, and organisational constraints in a group of mental, occupational and public health service users. A third of users, mostly women, described the working environment as negative for mental health. The main reported organisational constraints were poor career possibilities, relationship with the public and workload. Inappropriate workload is the leading cause for a negative evaluation of working conditions for mental health (Salerno et al., 2002a).
>
> Another Italian study reviewed literature on stress and women's work in order to discuss prevention priorities. Italian women are typically employed in the textile, clothing, shoes, food, pharmaceutical and tile sectors, in the teaching and nursing professions, and in services such as dry-cleaning, cleaning, hairdressing, etc. The study found that many jobs performed by women were characterised by monotonous and repetitive tasks that were simple but required concentration (this is also a characteristic of domestic work, whether paid or unpaid), hectic work and high demand, low-control jobs, and that women report more mental fatigue, dissatisfaction, bullying and low-paid work. A high prevalence of upper limb disorders was also found (Salerno et al., 2002b).

related to their role within the power structure of the organisation whereas female employees reported experiencing more severe stress when there was a conflict between job and home responsibilities.

More research is needed to identify the social and psychological processes through which supporting and caring for others on the job affect vulnerability to stress (Williams and Umberson, 2000).

> **Box 17**
>
> **Stress cycle: Swedish schoolteachers and stress**
>
> According to a Swedish survey of 1 000 pre-school teachers and lower elementary school teachers (Lärarförbundet, 2002), stress has increased in schools among teachers, pupils, and their parents. In all, 91 % of the respondents were women. The large number of pupils per class was identified as the main reason for teachers' high stress levels. Also, parents who come late to collect their children because of overtime work were seen as an important source of stress.
>
> Teaching is a good example of a profession that is emotionally demanding. In addition to the interactive nature of the work, teachers have to work in a noisy environment that per se causes stress. Prolonged standing may be harmful for reproductive health, and cardiovascular and musculoskeletal health, and teachers may suffer voice problems from the amount of talking they have to do, particularly when they have to raise their voice to speak loudly (Sala et al., 1999). In addition, teachers may receive high levels of criticism or blame from parents and pupils, for example for any classroom problems or poor exam results. Stress among teachers will not only affect their own health, but will also affect the quality of their work and their interactions with pupils.
>
> **Occupational stress and the parent–child relationship**
>
> According to Kinnunen et al. (2001), burnout, stress and other negative events in working life are reflected in the parent–child relationship. Briefly, a stressed parent pays less attention to her/his children, gives less positive feedback, and simply cares less. Children sense their parents' stress easily and respond to it immediately. It has been found that children have different expectations concerning their mother and father: it is the mother who is expected to provide care even in the case where she works longer hours than the father (Kinnunen, 2002).

Do types of stress and stress-related health outcomes vary between women and men?

While there is a consensus about what causes work-related stress, the reactions to the same psychosocial exposures vary between individuals. In other words, the stress-related symptoms or ill health experienced may vary among individuals. There are a number of reasons why it is important to look at stress reactions, for example to:
- recognise manifestations of work-related stress;
- accurately link health problems to the workplace;
- improve support for individuals experiencing stress.

As part of this, it is important to consider gender in relation to the experience of stress-related problems. In this report, we will look closer at burnout, depression and coronary heart disease. This includes exploring whether health problems are appropriately recognised as work-related for all workers.

Burnout and gender

Burnout consists of overall mental exhaustion, an increasingly cynical attitude towards work, and crumbling professional competence. The relationship between gender and burnout in general is unclear (Bakker et al., 2000). Although some studies or data sets show that women report more burnout than men, others show the opposite (Bakker et al., 2000; Houtman et al., 2000b; Taris et al., 2000). Kalimo and Toppinen (1997) found in their survey which was representative of the working-aged population in Finland (53 % of respondents were women) that burnout was only slightly more common among women than men. Of the burnout symptoms, exhaustion was found to be slightly more common among women than men, whereas cynicism was slightly more common among men than among women. Gender roles may explain this difference. The top five industries with the highest incidence of burnout were hotels and catering, banking and insurance, education and research (these three industries are female-dominated), mechanical repairs (male-dominated), agriculture and forestry (mixed) (Kalimo and Toppinen, 1997).

One explanation for the conflicting findings may be that many burnout studies have concentrated on specific occupational groups, and burnout has been shown to have a strong link with occupation. Schaufeli and Enzmann (1998) collected 73 studies, which showed quite large differences in burnout according to occupation, dependent upon the burnout subdimension that was analysed. It was shown that teachers — followed, at some distance, by health and social workers — were the most emotionally exhausted. Social workers, nurses, police officers and correction officers felt reduced personal accomplishment and police officers (followed by physicians) were the most distant and cynical. Managers and police officers reported quite low scores on emotional exhaustion. An analysis of 29 studies of burnout among different 'people work' occupations by Bakker et al. (2000) also found high risks of burnout among the 'caring for people' occupations, with general practitioners, homecare personnel, those in the psychiatric sector, nurses and those in the social care sector at the top end of the 'at risk' list and police work at the bottom end.

On the other hand, there is evidence that women may act in more positive ways than men to overcome burnout. According to Hakanen (1999), women have a wider range of supporting relationships than men, they take part in training sessions dealing with burnout, they take sick leave more often than men, and they tend to seek help for their problems from outside when necessary.

Not enough is known about the different reasons causing burnout in men and women. For

---

Box 18: Sick leave and burnout among young, professional women in Sweden

According to the Social Insurance Institution's statistics, young, educated women with a high salary lead burnout and sick leave in Sweden's statistics. Paid sick leave absences among women under 35 more than doubled between 1997 and 2001, with psychological reasons being the most common. Sjögren and Rappe (2002) investigated the reasons for this and found, among other factors, temporary, fixed-term contracts, inequality and poor management.

Deputising, temporary contracts and project work increase pressure to constantly prove one's skills. It may appear that individual workers have a lot of discretion within the job but this is often accompanied by a lack of clear goals, conflicting demands, etc. In addition, competent workers prepared to work long hours will often be assigned even more work. Women may feel under additional pressure to 'prove that they are up to the job', for example by putting in long hours, being very conscientious and setting very high work standards for themselves.

The statistics:

- In 2002, over 120 000 Swedes had been on sick leave for more than a year. Nearly a third of those on long-term sick leave were absent from work due to psychological reasons.
- Women take sick leave more often than men: women's share of sick leave in Sweden is 65 %.

example, Pretty et al. (1992) found in their study of male and female executives that pressures at work were causing work-related exhaustion. However, the pressures for women were related to interpersonal relations at the workplace whereas for men burnout was caused primarily by organisational factors. In addition, problems in private life caused extra pressures for women whereas men's burnout was strictly work-related, even when they were married and their wives also worked.

Depression

According to a recent WHO report and two well-known studies on the epidemiology of mental disorders, the epidemiological catchment area study (ECA) and the national comorbidity survey (NCS), women predominate in the mental disorders of depression, anxiety and somatic complaints. Another study (Middleton et al., 2001) suggests that women seek treatment for depression at twice the rate of men. Furthermore, research evidence suggests that depression is not only the most common mental problem among women but may also be more persistent among women than among men. In contrast, the lifetime prevalence rate for alcohol dependence, another common disorder, is more than twice as high in men than women in developed countries. Men are also over three times more likely to be diagnosed with antisocial personality disorder than women (Robins and Regier, 1991; Kessler et al., 1994; WHO, undated; Middleton et al., 2001; Haslam et al., 2003). In developed countries, gender-based roles, stressors and negative life events contribute to the morbidity (Ayuso-Mateos et al., 2001; Bildt and Michélsen 2002).

Gender-specific risk factors for common mental disorders that disproportionately affect women include gender-based violence, socioeconomic disadvantage, low income and income inequality, low or subordinate social status and rank and unremitting responsibility for the care of others. However, doctors are also more likely to diagnose depression in women compared with men, even when they have similar scores on standardised measures of depression or present with identical symptoms.

The role of work in the progress of depression needs further investigation. However, there is now consistent evidence from a number of cross-sectional and longitudinal studies that high levels of psychological demands, including high work pace and high conflicting demands, are predictive of poor mental health (Stansfeld, 2002). High job strain has been associated with higher rates of major depressive episode, depressive syndrome and dysphoria measured by the diagnostic interview schedule in the ECA study (Mausner-Dorsch and Eaton, 2000). A careful case-control study of healthcare staff found that although acute stressful situations and chronic difficulties outside work were important in anxiety and depressive disorders, there were also independent effects of 'conflict of work role' and 'lack of management support at work' (Weinberg and Creed, 2000). Saurel-Cubizolles et al. (2002) found that nursing staff reporting exposure to violence at work had a higher prevalence of depressive symptoms and stress symptoms. Haslam et al. (2003) suggest that high workloads, insensitive management, poor communication, low organisational awareness of mental health issues and poor work relationships are contributing factors to anxiety and depression. They found that mental health problems do not appear to be well understood by employers and managers, and little support may be offered in the workplace for individuals suffering anxiety or depression (Haslam et al., 2003).

Finally, it is unclear why there is such a big difference between men and women in the coping behaviour where mental problems are concerned: women tend to use medication prescribed by a doctor whereas men are more likely to turn to alcohol and narcotics. From the point of working life, both medication and substance use affect job performance, cause absenteeism, and increase the risk of having an accident. Also, returning to work from a long sick leave or rehabilitation scheme can be difficult (Hunter et al., 1998; Kivistö et al., 2001).

According to the WHO report, research shows that there are three main factors that are highly protective against the development of mental problems, especially depression. These are:
- having sufficient autonomy to exercise some control in response to severe events;
- access to some material resources that allow the possibility of making choices in the face of severe events;
- psychological support from family, friends, or health providers.

Haslam et al. (2003) propose that within the workplace: there is a need to raise awareness of mental health problems and increase understanding of how these conditions impact on working life; measures should be taken to prevent work-related stress; rehabilitation requires good coordination between managers, personnel staff, occupational health services, individual risk assessment and monitoring and the possibility of flexible working practices.

Coronary heart disease

Stress is one of the work-related factors linked to coronary heart disease. According to the EU labour force survey, men are more likely to report work-related heart disease and similar complaints as being their most serious complaint than women — 5.4 % of men compared to 2.4 % of women reported it as being the most serious complaint experienced (Dupré, 2002). Despite the fact that women do report work-related health complaints related to heart disease and given the importance of coronary heart disease as a cause of death among women as well as men, it is of concern that many studies of heart disease in relation to occupation have excluded women, or have looked only at male-dominated occupations.

NIOSH has summarised some of the available research on the links between coronary heart disease and occupation. It is known, for example, that certain occupational toxins, particularly carbon disulphide, nitroglycerine and carbon monoxide, affect the heart. Also, environmental tobacco smoke and extreme heat and cold are risk factors. Numerous studies show a link between exposure to work-related stress factors and heart disease. There is also evidence to sup-

Box 19: Stress and coronary heart disease

In a Swedish case study, women and men were compared with respect to sensitivity to psychosocial risk factors and coronary heart disease. Significant differences appeared between the sexes in areas of work content, workload and work control, physical stress reactions, emotional stress reactions and burnout. Results suggest that women are considerably more sensitive than men with respect to psychosocial risk factors for coronary heart diseases (Hallman et al., 2001).

Another study on female managers suggests that women unwind more slowly than men on return home from work. This finding can be explained by differences related to occupational level and/or sex, in regard to autonomy and social support at work, competitiveness, sex role, and reported conflict between paid work and domestic responsibilities. In the study, the stress profile of the female managers was considered in terms of possible long-term health risks, such as cardiovascular disease (Frankenhaeuser et al., 1989; see also Lundberg and Frankenhaeuser, 1999).

A Finnish study (Kivimäki et al., 2002) states that high job strain and effort–reward imbalance seem to increase the risk of cardiovascular mortality. The study was carried out in the metal industry among 812 employees (545 men, 267 women) who were free from cardiovascular disease at the baseline. The mean length of follow-up was 25.6 years.

The European Heart Network (1998) estimates that 16 % of cardiovascular disease cases among men and 22 % among women are due to work-related stress.

port an association between noise and elevated blood pressure. In addition, shift work, which disrupts body rhythms, has been linked to heart disease. Both underactivity and overexertion, particularly heavy lifting, are risk factors too (NIOSH web site (b)). Women can clearly also be exposed to work-related heart disease risk factors.

The results of a few recent large epidemiological studies suggest that psychosocial risks at work, particularly low control, have a link with a higher cardiovascular disease mortality risk for both men and women, even when variables such as age, cholesterol, obesity, and socioeconomic status are controlled (Alterman et al., 1994; Kornitzer et al., 2002; Kuper et al., 2002; Marmot et al., 1997; Bosma et al., 1997; Steenland et al., 1997).

Some studies have looked at psychosocial risk factors in coronary heart disease and gender differences in sensitivity and suggested that women may be more sensitive (Hallman et al., 2001; Wamala et al., 2000). The combination of high demands, low control and low reward is particularly harmful and many women are employed in this kind of 'routine work'. It is also known that shift work has an impact on hormonal levels and there is research evidence that, on average, female nurses doing shift work smoke more and are more overweight than day workers. These also contribute to health problems, including coronary heart disease (Kivimäki et al., 2001). In addition, women who have been prescribed hormone replacement therapy after the onset of the menopause and women who reach the menopause prematurely following a hysterectomy run an elevated risk of breast cancer and chronic heart and cardiovascular disease. This risk may be further increased because of the nature of work (Westerholm, 1998).

Large differences are seen in the risk of cardiovascular diseases among occupational groups.

Despite this, a Danish study found that 16 % of the cardiovascular disease among men and 22 % among women could be prevented by removing occupational risk factors from the work environment. The biggest risks include job strain, 6 % for men and 14 % for women, and shift work, which covers 7 % for both sexes (Kristensen et al., 1998).

Epidemiological studies on risk factors at work and cardiovascular disease mortality seldom concentrate on women. One reason for this is that women tend to get heart disease 10–20 years later than men; in practice, this means that most women are already retired when they fall ill and the occupational link thus remains unstudied. Compared to men, cardiovascular disease mortality is considerably lower among working women below the age of 50–55. Even when the focus is on women, the overall number of those who die before their pension has been too small to draw conclusions, as large cohort studies published on cardiovascular disease mortality in women are few. Low cardiovascular disease mortality has been explained by the fact that the female hormones — progesterone and oestrogen — protect women. However, this protection is lost after the menopause (Alterman et al., 1994; Kornitzer et al., 2002; Kuper et al., 2002; Marmot et al., 1997; Steenland et al., 1997; Houtman and Kornitzer, 1999; WHO, 1994b).

Clearly, further research into occupational risks of coronary heart disease is warranted, and this research should be gender-sensitive.

Prevention of stress and stress-related diseases

The European Agency report *Research on work-related stress* (Cox et al., 2000) concluded on the basis of available scientific evidence that work-related stress can be dealt with in the same way as other health and safety issues, by adapting the control cycle already well

established for the assessment and management of physical risks to the management of stress at work. A second Agency report (European Agency, 2002a), based on an analysis of examples of workplace interventions and tools to prevent work-related stress, presents key success factors for stress prevention in the workplace. These are:
- base the intervention on an adequate risk analysis;
- thorough planning and stepwise approach — including setting clear objectives and target groups;
- combination of work-directed and worker-directed measures, with priority given to collective and organisation interventions to tackle risks at source;
- context-specific solutions, with employees' on-the-job experience being a vital resource in identifying problems and solutions;
- when outside expertise is used, using experienced practitioners and evidence-based interventions;
- social dialogue, partnership and workers' involvement;
- sustained prevention and top management support.

Therefore, as a starting point, it is very important to look at the real tasks and activities that women and men are performing and the real conditions and circumstances of their working in order to take account of all the factors that could be contributing to work-related stress. Men's and women's working conditions and experiences are not identical and so they will face different exposures to the same stressors and exposure to different stressors even in the same workplace and in the same job. It is also essential to ensure the effective participation of women in the process. It is important to understand the role of gender when planning stress-prevention programmes as, due to job segregation, men and women become stressed for different reasons in the same workplace. So far, many interventions that have not been sensitive to gender have not always produced effective results (for example, see Östlin, 2002). Risk assessment should therefore take account of stressors such as sexual harassment, discrimination and family responsibilities, as well as those found more often in female-dominated jobs. This could include, for example, looking at individual control over work and unforeseen interruptions.

Many employers are recognising the need to take steps to prevent work-related stress. There are examples of good practices to prevent stress in female-dominated work areas in three Agency reports (European Agency, 2002a, 2002b, 2002e). The cases show that it is possible to intervene and make improvements in the emotionally demanding work of healthcare workers and teachers, give shift workers more control over their working hours rotas, and introduce more autonomy, participation and control into the jobs of care assistants and call centre operators, for example. Interventions that can be made to improve work–life balance, including for those with families and caring responsibilities, are covered in the section on work–life balance.

Coping

The priority in work-related stress is to prevent the sources of stress in the first place. However, coping is an overall part of the stress process. It is an attempt to manage demands and can be viewed as a problem-solving strategy. The styles and strategies used need to be relevant to the particular situation. However, it is probably the least well-understood area of work-related stress research and the need for more information on coping is widely recognised (Cox et al., 2000).

Houtman and Van den Heuvel (2001) examined international studies on sex and gender differences in physical fitness, personality, attitudes

> **Box 20: An example of a coping intervention in a government department in the UK**
>
> A stress-management programme was carried out in a government department employing over 25 000 workers. The organisation had just gone through a structural change and job satisfaction had decreased, while stress levels had increased. In their intervention, Whatmore et al. (1999) aimed to improve coping among individual employees. In all, 270 (157 of them women) staff members participated in the study. Three areas were investigated: education and awareness, exercise, and cognitive restructuring. Benefits were seen for individuals engaged in the exercise programmes and in the awareness training after three months, and this effect was still seen with the exercise programmes after six months. However, the cognitive restructuring programme did not have a positive effect. Organisational variables, i.e. job satisfaction and organisational commitment, were not affected by the interventions. The self-reported sickness absenteeism rate was lowered in the exercise group but increased in other groups. In conclusion, the researchers state that individually targeted stress-management interventions may produce benefits, but for the maintenance of these benefits, identification and minimisation of organisational stressors are required.
>
> Source: Whatmore et al. (1999).

and aspects of coping. This included looking at gender differences in coping with job demands. The majority of the research reviewed only looked at gender differences in coping capacity in relation to health outcomes. Few studies were found that explored the moderating or mediating effects of indicators of coping capacity in the risk–outcome relationship.

In a study of occupational stress in female and male managers, Davidson et al. (1995) found that females were more likely than males to adopt positive coping strategies, for example they made significantly greater use of social support and task strategies.

Given the demographic changes in the workforce, the ability of older workers to cope with work has begun to be discussed, particularly with respect to changes in the workplace, such as the introduction of new technologies or new work methods. This discussion is in relation to all aspects of work, not just stress. One review in this area found that older workers tend to be thought of as a homogeneous group, and little attention has been given to any possible gender differences. However, for example, an older woman worker is more likely to have caring responsibilities for a disabled parent or partner. The review found that older women workers (aged over 45), in particular, are neglected both in research and in policy areas (Doyal, 2002). Older women will be discussed again in a later section.

Why don't women suffer more work-related stress?

Given that many of the known causes of stress are features of female-dominated jobs and the additional pressures of work in the home, it may be worth asking the question 'Why do women not suffer more from stress?'. Women's better and more frequent use of social support and more task-oriented coping mechanisms have been mentioned above. Research evidence also suggests that the physical and mental health of working women, with or without children, is greater than that of non-working women (for example, Rout et al., 1997). Married working women with children are the healthiest of all women (for example, Kivelä and Lahelma, 2000). In other words, working life may increase stress but it also offers social contacts and other benefits that are not available at home. Paid work has positive emotional and social implications that contribute to well-being, and it also positively affects economic well-being (Bildt Thorbjörnsson and Lindelöw, 1998).

Improving research and future research priorities

The European Agency report *Research on work-related stress* identified a number of areas for further research, including on stress interventions and management at enterprise level; new patterns of work such as teleworking; and coping ability. In addition, the report concluded that there is a need to place stress at work in the context of other interrelated issues such as gender and that this is essential in order to achieve a complete perspective for the management of stress at work (Cox et al., 2000).

Although many studies analyse gender as a variable and most survey instruments include scales that examine stress factors and moderators such as job demands and control, and relationships with others at work, few include working conditions that are more common for women such as working conditions incompatible with family responsibilities, sexual harassment, discrimination, and invisible work (Messing, 2000). There is some evidence that gender differences exist even in these 'classic' factors; for example, studies have found social support to be a more potent stress buffer/moderator for women than for men (American Psychological Association, 1996). These gender differences and their implications deserve further exploration. Also, the prevalence and the health impact of stressors specific to women and ethnic/minority groups have not been examined sufficiently.

Again, as with other health issues, research is hindered where conditions are either not recognised as work-related, so data are not collected, or there is disagreement on classification. For example, in the Netherlands, a classification system has been introduced (CAS-coding), which includes burnout as a subcategory of mental disorders. However, there is no agreement on the classification of a mental health problem (burnout or depression) among occupational health physicians and general practitioners (Houtman et al., 2002).

Injuries and stress arising from work-related violence from members of the public

Key points

- Both men and women experience work-related violence. Women are more exposed because exposure is strongly associated with 'people' work.
- Those affected go far beyond 'victims', including witnesses, and others who experience fear in the job.
- Effects are wide ranging from death and physical injury through to stress.
- The steps organisations can take to prevent work-related violence are well documented. More research and prevention good practice by sector are needed, as well as sharing of existing information.
- Not all countries include violent incidents in official work-related accident figures.
- In data collection, a clear distinction should be made between incidents of violence from members of the public and workplace bullying.

The European Commission defines violence arising out of work as 'incidents where persons are abused, threatened or assaulted in circumstances related to their work, involving an explicit or implicit challenge to their safety, well-being or health'. This definition covers both verbal and physical abuse, such as physical attacks, robbery, threats, shouting and verbal abuse (Wynne et al., 1997). As well as being a cause of physical injury, work-related violence is a major source of stress. In comparison with men, women also suffer more from post-traumatic stress following a violent incident, for example a robbery (Yehuda, 2002).

According to the third European survey on working conditions, 4 % of the working population report that they have been victims of actual physical violence from people outside the workplace in the previous 12 months, with women reporting higher levels (Paoli and Merllié, 2001; see also Cooper et al., 2003). Many more will have suffered from threats and insults or other forms of psychological aggression. For example, according to a Swedish survey, 17 % of women and 9 % of men had been exposed to violence or the threat of it at the workplace in 1997 (cited in Menckel, 2001). The abuse or intimidation may include sexual or racial harassment. Fear of violence is widespread and is a problem in its own right, particularly in high-risk occupations, and there will be negative consequences for those who witness violent incidences (Cooper et al., 2003). This increases still further the size of the workforce affected. Fagan and Burchell (2002) state that the figures in the third European survey on working conditions are 'certainly an underestimation of the true rates of intimidation, harassment and discrimination given the limited nature of the question asked in the survey'.

According to Cooper et al. (2003), a distinction also needs to be made between the various forms of violence since 'women have been found to be over-represented among victims of intimidation and psychological violence (while men are more frequently exposed to physical violence and assault). Potential differences between men and women in their willingness to accept and label their experience as violence and harassment will impact on these findings.'

However, injury and even death to women workers can be the result of violent incidents in the workplace. In the United States, homicide is the leading cause of death for women in the workplace accounting for 40 % of all workplace deaths among female workers. These workplace homicides are primarily robbery-related, and often occur in grocery/convenience stores, eating and drinking establishments, and gasoline service stations (NIOSH, 2001). In the United States, the level of this violent crime will be related to the greater use of guns in violent crime, compared to EU countries. In the EU, there have been incidents of healthcare workers killed by psychiatric patients, a social service housing officer killed by a client, and incidents of angry and disturbed people entering schools with firearms.

In some EU countries, incidents of workplace violence fall outside the scope of statutory reportable incidents and are, therefore, not registered at a national level (Cooper et al., 2003), resulting in an underestimation of work-related injuries in the national statistics, as well as a lack of information for preventive action. For example, Saurel-Cubizolles et al. (2002) carried out a study in the healthcare sector, promoted by knowledge that there was a significant problem and a lack of data about the real extent of the problem (see also Semat, cited in Vogel, 2002). Also, some studies and data collection activities do not clearly separate incidents related to violence from members of the public from workplace bullying perpetrated by work colleagues, although they are two separate issues.

> **Box 21: Policewomen in the firing line**
>
> According to a study of Finnish policewomen in 1998, 24 % had experienced physical violence at work and 22 % had been injured in an attack. Nearly half of the policewomen had been threatened face-to-face and 21 % on the telephone. The policewomen had experienced verbal sexual harassment — such as sexist jokes and offensive comments — from clients as well as from work colleagues. The research results suggest that violence and sexual harassment can be linked to job stress and burnout.
>
> *Source:* Kauppinen and Patoluoto (in press).

Women's increased risk of violence from members of the public is again at least partly related to job segregation, where women are often concentrated in high-risk jobs and occupations with respect to violence such as nursing, social work and teaching. Workplace violence is widespread across all occupations where workers come into contact with members of the public. In the third European survey on working conditions, 64 % of women compared to 48 % of men reported that their work involved contact with customers, pupils, patients, etc., for at least half of the time (Fagan and Burchell, 2002). Some high-risk occupations and risk factors are given in Box 22.

Women's jobs are generally characterised as 'people work'. Women in managerial and professional jobs are also found to do more 'people work' than men employed in these occupational categories. A good example is the work of a nurse: while a typical professional female nurse works directly with the patient in the sick ward, her male colleague is more likely to work in the intensive care unit, the operating room, or is the head of the department. In other words, a typical male nurse has less patient contact than a female nurse (Dassen et al., 1990).

Younger people are also more at risk than older people (Cooper et al., 2003). For example, in the UK 1995 labour force survey, 8 % of women workers reported being physically attacked by a member of the public in the course of their work — rising to more than one in ten 25–34-year-olds — and nearly one in five women workers had been threatened with physical violence. In both cases, the rates for women were about 20–30 % higher than for men. This age effect may be due to job segregation, with younger people having more direct contact with members of the public, possibly coupled with inexperience as well as lack of authority in the job, for example to resolve a client's problem.

It should also be noted that there are large country differences in reports of work-related violence. Country differences in the social and economic context, such as general levels of crime, economic and social change, the presence and, in some countries, growth of the informal economic sector could have an impact on types and levels of violence (Cooper et al., 2003).

Severe under-reporting of incidents was found in an Italian study in the psychiatric care sector,

---

Box 22: Work-related violence: where are the risks, what are the consequences?

Risky environments include:
- Service sector, and particularly organisations in the health, transport, retail, catering, financial and education sectors.

Occupations particularly at risk include:
- Nurses and other healthcare workers
- Bus drivers
- Cashiers (including in banks and shops)
- Parking inspectors
- Social workers and others providing social services
- Restaurant and bar staff

Risk factors include:
- Contact with customers or clients
- 'Enforcement' activities
- Working alone or with a few workmates
- Handling money

The consequences for the worker vary greatly, ranging from:
- Demotivation and reduced pride in performing one's job
- Stress (even for the indirect victim, the witness of the violent act or incident)
- Injury or even death
- Physical or psychological ill health
- Post-traumatic symptoms like fear, phobias and sleeping difficulties
- Post-traumatic stress disorder, in extreme cases

Based on information from: European Agency (2002c); *Current Intelligence Bulletin* (1996); Cooper et al. (2003).

despite the effect of assaults on the physical and psychological health of staff (Magnavits, 2002). In some work, violence may be seen as 'part of the job' or there may be reluctance to blame the perpetrator. There will also be under-reporting if victims feel they will be blamed, if there is unlikely to be any preventive action taken, or if there is no system for reporting. The study found that female nursing staff were in the front line, and incidents included scratching, spitting, slapping, kicking and punching.

The gender of the aggressor in a violent incident may play a role in encouraging or discouraging violence, depending on various circumstances. Some research evidence suggests that, for example, policewomen may be less authoritarian and use force less often than their male counterparts, are better at defusing potentially violent confrontations, possess better communication skills, and respond more effectively to incidents of violence against women (National Centre for Women and Policing web site). The presence of a woman may also have an 'educating' influence on the behaviour of some potential assailants, and there is some anecdotal evidence regarding all male and mixed-gender ambulance crews to support this, with the inclusion of a woman in the crew appearing to have a calming influence. On the other hand, female police officers may provoke aggression in some situations and some research suggests that women police officers would prefer to attend a difficult situation paired with a man rather than another woman (Nuutinen et al. 1999).

The steps to successful prevention of work-related violence are well documented. There is no specific EU-level legislation in this area, although it is covered by the general requirements of risk assessment and prevention in the EU legislative framework and individual Member States have taken a variety of individual measures.

Further research and development of good practice in specific sectors, occupations and types of violence is strongly recommended (Cooper et al., 2003).

Box 23
International action on violence in the health sector

Violence in the health sector is a major problem worldwide, affecting over 50 % of health workers. Ambulance and pre-hospital emergency staff are at greatest risk from violent attacks. Nurses are three times more likely to experience violence at the workplace than other occupational groups. While both men and women are affected by workplace violence in the health sector, the majority of the workforce is female. Sources of violence include patients under the influence of alcohol or drugs, for example in emergency clinics, patients whose behaviour is affected by the drugs they are taking or their illness or injury, or patients or relatives who are angry or frustrated. Large hospitals in suburban, densely populated or high-crime areas, as well as those located in isolated areas, are particularly at risk. In many countries, reporting procedures are lacking and perpetrators are not prosecuted. In response, a joint task force on workplace violence in the healthcare sector, comprising the ILO, the WHO, Public Services International (PSI) and the International Council of Nurses (ICN), launched a set of 'Framework guidelines for addressing workplace violence in the health sector', based on case studies including ones from Portugal and Bulgaria and other Member State and worldwide experiences. The guidelines are intended to support everyone who is responsible for safety in the workplace, and to guide initiatives at international, national and local levels. The ILO is also expected to adopt a 'Code of practice on violence and stress at work in services — A threat to productivity and decent work' in October 2003.

Sources: ILO et al. (2002a); ILO (2002b); Cooper and Swanson (2002); di Martino (2002).

A hospital acts to prevent violence to staff

A survey showed that in a Dutch hospital most incidents were occurring at the reception/switchboard, and in the accident and emergency and psychiatry departments, and during weekends, evening and night shifts. A working party, involving staff and managers, investigated the measures needed. A plan of action was agreed with the works council and executive board. This included issuing staff with alarms to be used to alert security staff and setting up a close cooperation with the police which included provisions to call them in to assist in incidents when necessary and for them to provide information to hospital staff about potential problem patients. A system of issuing yellow and red cards to aggressors was introduced to inform them when their behaviour was not acceptable and in severe cases banning them from entering the hospital. Posters and leaflets describing the policy were placed in clinics, community centres and other public places and notices were put in the local press, to make the public aware of the policy. Initiatives such as display boards to keep patients informed in waiting areas were also introduced. The measures were backed up with staff training and provisions for counselling, and regular discussions of the programme at staff meetings.

Source: European Agency (2002b).

The effects of sexual harassment from colleagues

Key points

- Sexual harassment is a workplace stressor experienced much more frequently by women than by men.
- Sharing and exchange of current legislation, policies and good practice will support the implementation of the amending Council directive covering sexual harassment at work.

Sexual harassment is increasingly acknowledged as an affront to dignity at work (Cooper et al., 2003; Herbert, 1999; Kauppinen, 2003). The European Commission's code of practice defines sexual harassment as 'unwanted conduct of a sexual nature, or other conduct based on sex affecting the dignity of women and men at work'. The definition includes: unwanted, improper or offensive behaviour; behaviour that influences decisions concerning jobs; behaviour that creates a working environment that is intimidating, hostile or humiliating. Forms of sexual harassment include: verbal, e.g. sexual jokes; non-verbal, e.g. staring; and physical, e.g. unsolicited touching (European Commission, 1991; European Commission, Employment and Social Affairs, 1998).

According to the third European survey on working conditions, sexual harassment was experienced by 2 % of respondents in the previous 12 months. This figure is double in the Nordic countries which probably relates to greater awareness. Women are subjected to these issues to a much greater extent than men (4 % compared to 2 %) and the rate is higher for temporary agency workers than for employed workers (Paoli and Merllié, 2001).

European studies estimate that approximately 30 to 50 % of women have experienced some form of sexual harassment or unwanted sexual behaviour in the workplace compared to about 10 % of men. Men also perceive harassment as less offensive and they experience less negative consequences (European Commission, Employment and Social Affairs, 2002a). Younger women and women in male-dominated workplaces experience more sexual harassment (Kauppinen, 1999b).

Despite being commonplace, sexual offences at work go unreported in many cases due to factors such as fear of losing one's job, being considered the guilty party or being socially ostracised among work colleagues (Cooper et al., 2003). See also the comment in the section on violence about the underestimation of the problem in the third European survey on working conditions.

Sexual harassment can have negative outcomes for both the individual exposed and the organisation. It is a recognised workplace stressor and affected employees may report negative effects on their health and well-being and on their job motivation and satisfaction. Their career advancement may also suffer. It may affect their work performance and they may take sick leave. Sexual harassment can also have a demoralising effect on the general atmosphere at work (Kauppinen, 1999b). Employees experience less harassment and greater job satisfaction in organisations that are characterised by a positive and open social climate (Herbert, 1999).

It is recognised that organisations should have preventive policies and procedures to deal with sexual harassment (Bleijenbergh et al., 1999; Herbert, 1999; Kauppinen, 2003). Risk assessments for stress should include sexual harassment. Key elements that policies should cover are given in Box 24. The amended European directive on equal treatment (Council Directive 2002/73/EC amending Council Directive 76/207/EEC) deems sexual harassment at work to be discrimination and therefore prohibited, and requires Member States to take measures to prevent it. It comes into force in 2005. If fully implemented across the Member States, it would go a long way to helping to prevent an important workplace stressor for women. There are various examples of legal approaches and practices aimed at prevention. For example, Kauppinen (1999b) describes how sexual harassment can be included in workplace equality plans. Many examples of guidelines to prevent sexual harassment exist. Guidelines tailored to particular types of work can be useful, and a guide aimed at preventing sexual harassment in sports is one such example (Tiihonen and Koivisto, 2002). Information exchange on current practices both within and outside the EU would support the implementation of the directive.

The effects of workplace bullying

Key points

- Women are more likely to suffer bullying at work. Gender segregation — the concentration of women in the public sector, where bullying is more prevalent — is likely to be one reason for this.
- Developing and sharing of Member State guidelines, good practice and training materials would be helpful for prevention, as well as for developing a common understanding among Member States.
- The European survey on working conditions does not clearly distinguish between intimidation from members of the public and from colleagues. It would be helpful if the survey asked separately about both sources as well as exploring the issues of violence, bullying and sexual harassment in more detail.

Bullying, mobbing, and psychological or moral harassment are various terms used to describe intimidating behaviour between colleagues in the workplace.

The third European survey on working conditions suggests that women are more likely to experience intimidation and bullying at work than men (10 % as opposed to 7 % reported experiencing it in the previous 12 months). These figures are a slight increase on the 1995 survey. There are wide differences between

---

Box 24: Tackling sexual harassment: key elements for workplace policies

- A clear, contextual definition of sexual harassment.
- Detailed provisions for preventive measures.
- A complaint procedure and complaint officer.
- Protection and support for harassed employees.
- Sanctions for those found guilty of harassment.
- Supportive initiatives such as special training programmes, designed to raise awareness of the issue and to equip those with responsibility for carrying out the procedures.

Source: Bleijenbergh et al. (1999).

> **Box 25**
>
> **Workplace bullying can be defined as**: repeated, unreasonable behaviour directed towards an employee, or group of employees, that creates a risk to health and safety.
>
> Within this definition:
>
> - 'unreasonable behaviour' means behaviour that a reasonable person, having regard to all the circumstances, would expect to victimise, humiliate, undermine or threaten;
> - 'behaviour' includes actions of individuals or a group. A system of work may be used as a means of victimising, humiliating, undermining or threatening; 'risk to health and safety' includes risk to the mental or physical health of the employee. Bullying often involves a misuse or abuse of power, where the targets can experience difficulties in defending themselves.
>
> **The consequences of bullying** may be significant for the victim.
>
> Physical, mental and psychosomatic health symptoms are well established, for example stress, depression, reduced self-esteem, self-blame, phobias, sleep disturbances, and digestive and musculoskeletal problems. Post-traumatic stress disorder, similar to symptoms exhibited after other traumatic experiences such as disasters and assaults, is also common among victims of bullying. These symptoms might persist for years after the incidents. Other consequences might be social isolation, family problems and financial problems due to absence or discharge from work.
>
> *Source:* European Agency (2002d).

countries — ranging from 15 % of workers being subject to intimidation in Finland to 4 % in Portugal — which probably reflects awareness of the issue rather than reality (see also the comment in the section on violence about the underestimation of the problem in the third European survey on working conditions). According to other surveys, approximately 5 to 10 % of the population perceive themselves as being bullied at any one time and a substantially higher number will be regularly exposed to behaviour that may be described as bullying behaviour without necessarily feeling victimised (Cooper et al., 2003).

Job segregation may, again, be a reason for women overall being more exposed, as there appears to be a higher risk of bullying and intimidation in the female-dominated public sector compared to the private sector (Cooper et al., 2003). In many individual studies, which are often confined to one type of sector or workplace, once account has been taken of the gender composition of the workforce, women do not appear to be more exposed than men (for example, see di Martino et al., 2002; Vartia, 2003).

Some surveys have shown that men are more likely to be the perpetrators of bullying, but that this is likely to be because they are more often in managerial positions (for example, Unison, 1997). One major survey in Ireland found that the perpetrators of bullying were a single supervisor or manager in 45.3 % of cases (Health and Safety Agency, 2001). Bullying at work is an abuse of power and one important source of bullying is managers who bully subordinates. This includes senior managers bullying more junior mangers. Both men and women may bully and both may be bullied.

There is general agreement that factors giving rise to workplace bullying relate to workplace culture and work organisational issues. For example, an analysis of data from the third European survey on working conditions (Daubas-Letourneux and Thébaud-Mony, 2002) suggests that high levels of bullying among both men and women are more prevalent where employees are expected to work in highly flexible work situations, characterised by, for example, a client or profit-driven work pace and demanding quality control procedures (see also Piispa and Saarela, 1999; Menckel, 2001). Bullying has also been found to be typical of large, poorly run hierarchical organisations with high work pressures, such as healthcare (Sutela, 1999). However, the precise bullying behaviour used to coerce or undermine (comments etc.) may be targeted at particular characteristics of the bullied person, so may be gender-related, or may depend on the type of work the bullied person does — for example, bullied people in managerial positions, usually men, more often report being given impossible deadlines than workers (Vartia, 2003).

As mentioned above, surveys suggest that there is a higher incidence of bullying in the female-dominated sectors of public administration, education, healthcare and social work as well as financial services sectors. There is also some suggestion that bullying is also linked to organisational change and these sectors may not only exhibit the characteristics described above, but often have also been subjected to corporate change in recent years.

Exposure to bullying is associated with high levels of stress (Daubas-Letourneux and Thébaud-Mony, 2002). Bullying affects both physical and psychosocial well-being, increases sickness absenteeism and may lead to a change of job. Suicidal thoughts are quite common in the most serious cases (Cooper et al., 2003).

European action in this area will reduce a source of workplace stress that affects women more than men. The European Commission's guidelines on work-related stress (European Commission, 2001) refer to bullying as a source of stress. The European Parliament has called on the Member States to review and, if appropriate, to supplement their existing legislation and to review and standardise the definition of bullying, with a view to counteracting bullying and sexual harassment at work. While few Member States have specific legislation on bullying, others are looking at bringing in legislation and some countries have taken regulatory steps by means of charters, guidelines, and resolutions. Development and sharing of Member State guidelines, good practice and training materials would be helpful for orienting action, as well as for developing a common understanding between the Member States (Cooper et al., 2003).

Box 26

Key elements of effective workplace policies to prevent negative social interactions include:

- commitment from employer and employees to foster an environment free from bullying;
- outlining which kinds of actions are acceptable and which are not;
- stating the consequences of breaking the organisational standards and values, and the sanctions involved;
- indicating where and how victims can get help;
- commitment to ensure 'reprisal-free' complaining;
- explaining procedure for making a complaint;
- clarifying the role of manager, supervisor, contact/support colleague, and trade union representatives;
- details of counselling and support services available for victim and perpetrator;
- maintaining confidentiality.

Source: European Agency (2002d).

Work and domestic violence

The issue of domestic violence is outside the scope of this report. However, women are more exposed than men and the issue can spill over into the workplace. The recipients may suffer either physical or psychological health problems and this may affect their work performance and result in their taking time off work. In addition, the threatening partner may, for example, make phone calls, send e-mails and text messages, and even visit the victim at the workplace. This will not only affect the performance of tasks but the whole atmosphere of the workplace (European Commission web site (a); End abuse web site; Miller and Downs, 2000; Valls-Llobet et al., 1999).

Infectious diseases

Key points

- Women are more affected by work-related infectious diseases than men.
- In work areas such as the healthcare sector, which are high-employment areas for women, there is a higher risk of exposure to infectious diseases.
- Minor infections contracted through work that cause sickness absence, for example by those working with young children, are not generally recognised as being work-related or recorded.

According to the EU labour force survey, women are more affected than men in most Member States by work-related infectious diseases (except Denmark, Spain and Portugal). Infectious diseases rank as the fifth most serious work-related complaint experienced by women and the eighth most serious complaint experienced by men (Dupré, 2002).

Workers in healthcare, schools and nurseries, as well as other jobs bringing them into contact with members of the public, may be exposed to infectious diseases as a consequence. Some infections can be transmitted from animals to humans. Those at risk of exposure include agricultural workers, veterinary workers and those working with laboratory animals. Women's segregation into the care work and other service sectors, in particular, means that they are more exposed to infectious diseases at work. For example, a study of the healthcare sector in nine Member States identified biological agents as a main risk factor (Verschenuren et al., 1995).

Biological hazards for health and social care workers include a whole range of minor to potentially life-threatening infectious diseases.

They may come into contact with infected blood or body fluids, including faeces and urine, as well as airborne infections. Serious bloodborne infections include hepatitis B and hepatitis C as well as HIV, although so far only relatively few cases of work-related HIV transmission have been registered (Box 27). Accidents with infected hypodermic needles or other contaminated 'sharps' are one form of transmission of bloodborne infections. These needlestick injuries are emotionally as well as physically threatening, as they can involve possible exposure to serious and life-threatening infections. Faecal-borne infections include hepatitis A and dysentery. Airborne infections include tuberculosis. It is not just nursing staff who are at risk from biological agents. So too are the whole range of different professions and support staff delivering direct care to patients, as well as others such as cleaners, laundry workers, and those working in laboratories or dealing with clinical waste.

In care work, scabies is common too: for example, in Finland in the year 2001, there were 109 reported occupational scabies cases among female healthcare workers and only eight among men (Ammattitaudit, 2001). The uneven distribution of this disease among men and women reflects, once again, the different types of jobs performed, even within one sector.

Those working with small children are more exposed to common colds, stomach infections, and various infections associated with childhood such as chicken pox, which can be more serious in adults.

In addition to the health effects on the worker, some diseases can affect the foetus of pregnant workers, such as rubella.

Especially more minor infections that may be contracted at work, for example by teachers or nursery nurses, are not usually officially recognised as work-related illnesses, so do not appear in work-related absence figures.

> **Box 27: Healthcare worker exposure to bloodborne diseases**
>
> Among the 35 million healthcare workers worldwide, about 3 million receive percutaneous exposures to bloodborne pathogens each year, 2 million of those to hepatitis B virus (HBV), 0.9 million to hepatitis C virus (HCV) and 170 000 to HIV. These injuries may result in 15 000 HCV, 70 000 HBV and 500 HIV infections. More than 90 % of these infections occur in developing countries. Worldwide, about 40 % of HBV and HCV infections and 2.5 % of HIV infections in healthcare workers are attributable to occupational sharps injuries. These diseases are largely preventable through good management, staff training, HBV immunisation, post-exposure prophylaxis and good waste management.
>
> *Source:* WHO (2002b), *World Health Report 2002.*

Asthma, other respiratory disorders, and sick-building syndrome

Key points

- Allergy symptoms are increasing in the general population and many are caused or made worse by exposure to sensitising agents at work.
- Across Europe, men seem to be more affected than women by work-related breathing and lung problems. However, in some individual Member States, the situation is reversed. There are many cases of work-related asthma among both women and men.
- Women-dominated work areas where there may be exposure to the most common causes of work-related asthma include healthcare work, textile manufacturing, food production and hairdressing.
- Women seem to experience sick-building syndrome symptoms more often than men, which could be due to different working conditions even when working in the same workplace.
- Reporting and recording of respiratory diseases vary greatly among Member States, making it difficult to assess the true extent of the problem as well as gender differences.

Asthma and allergies seem to be increasing in the industrialised world. Factors suggested as contributing to this problem include air pollution, additives, smoking, exposure to environmental tobacco smoke, and changes in diagnostic practice, just to mention a few. Many cases of asthma and other respiratory diseases in adults are caused or made worse by work. In addition to asthma, other occupational respiratory disorders include: rhinitis, organic dust toxic syndrome and allergic alveolitis.

The most common causes of occupational asthma are: isocyanates; flour dust; grain dust; glutaraldehyde — a powerful disinfectant used in hospitals; wood dust; natural rubber latex — for example used to make the fine protective gloves used by healthcare workers; solder/colophony; laboratory animals; glues and resin (Health and Safety Executive, 2002a). Women workers in the manufacturing, food and healthcare sectors are among those who may be exposed to these substances.

Agriculture and forestry and the sawmill industry have particularly high rates of respiratory diseases. However, the female-dominated industries of food processing and textiles also have high rates and occupational asthma is well documented among hairdressers due to the use of various sprays and chemicals at work (for example, Ammattitaudit, 2001). Many industrial cleaning agents used by cleaners can also cause asthma. Other factors that may contribute to gender differences include smoking habits, access to healthcare and biological differences.

In the EU labour force survey, 8.4 % of men and 6.4 % of women reported breathing or lung problems as being their most serious work-related health complaint. However, in Luxembourg, Portugal and the UK, the figures were higher for women (Dupré, 2002). In the third

European survey on working conditions 2000, 10 % of men and 8 % of women reported work-related allergy problems (these figures include both skin and respiratory problems), but this percentage difference is reduced to 1 % when the figures are compared across either full-time or part-time workers. For both male and female workers, the reported problems drop among part-timers, suggesting that the incidence of work-related allergies is related to exposure.

A study to look at work-related factors regarding the onset of asthma using recorded data of 25–59-year-old Finnish workers during 1986–98 found that 29 % of the new male asthma cases and 17 % of the new female asthma cases were attributable to work-related factors. These figures are in contrast to studies that have looked only at cases that were known to be due to specific allergens as defined by the national compensation scheme — which suggests that work-related asthma accounts for 6 % of adult onset cases in men and 4 % in women (Karjalainen et al., 2001). However, according to Finnish statistics from 2001, new female cases of allergic rhinitis and asthma exceeded new male cases by 301 to 212.

However, conclusions cannot be drawn solely on the basis of these figures. As the reporting and recording of respiratory diseases, particularly asthma, vary greatly among the Member States, it is difficult to assess the extent of this problem, including a gender-related picture (Meding and Torén, 2001). It is important to harmonise the collection of statistics and make the reporting of new occupational asthma cases compulsory.

Air pollutants are not just an issue for 'dusty' or 'fumy' workplaces. They are also an issue in offices, where a high proportion of workers are women. The issue of indoor air pollution in offices began to be discussed more seriously at the beginning of the 1980s, when sick-building syndrome began to be reported in buildings whose main source of ventilation was an air-conditioning system. Although there is no precise scientific explanation, it has been associated with various indoor air quality factors, such as chemicals, including in construction and furnishing materials, smoking residues and micro-organisms — bacteria and moulds — in the filters of ventilation systems. Poor indoor air quality may cause a variety of symptoms, such as headaches, watering eyes, nasal symptoms, cough, frequent colds, and sensory irritation in general.

Women report sick-building syndrome symptoms more than men. However, according to a Swedish study, job segregation and differences in work environment may at least partly explain this. For example, incidents of sick-building syndrome have been more common in public sector buildings and associated jobs, where a higher percentage of women are employed. This suggests that differences in working conditions, and not differences in sensitivity to symptoms, account for the increased incidence of symptoms among women (Stenberg and Wall, 1995; Bullinger et al., 1999; Jaakkola et al., 2000; Messing, 1998). In offices, women are more likely to carry out photocopying tasks and so are more exposed to toners, ozone and electrostatic effects (Messing, 1998). In addition, a number of researchers have pointed out that work-related stress should also be considered as a possible contributing factor (for example, Stenberg and Wall, 1995).

Multiple chemical syndrome is a condition that may also be related to sick-building syndrome and the symptoms are similar. It is also much more common among women (Messing, 1998; Kipen et al., 1995). Symptoms include respiratory and digestive problems, but can also include sensitivity to noise. It is thought to be triggered by prior exposure to organic solvents or other chemicals although some claim that it is of psychological origin caused, for example, by stress (Maschewsky, 2002; Environmental Law Centre web site; see also Fernandez et al., 1999; Black et al., 2000).

The number of employees, and particularly the number of women working in offices and similar kinds of environments, is increasing and this underlines the importance of taking seriously symptoms reported by workers in these environments, especially as the symptoms are likely to be due to a very large number of factors whose individual levels are not considered to be dangerous. Research focused on the link between indoor air and health outcome should adequately address gender issues and also properly take into account the working conditions that could be causing or contributing to these symptoms.

Skin diseases

Key points

- Women workers are more often exposed to detergents, solvents and water. 'Wet jobs' are particularly associated with hand dermatitis.
- Women have an increased risk of dermatitis in jobs such as electro-manufacturing work, hairdressing, healthcare work, mechanics and metalwork. There is also a high rate among kitchen workers and cleaners.
- Skin rashes are among the symptoms of sick-building syndrome, which appears to affect women more than men.
- VDU work is sometimes associated with skin problems such as itching and rashes — possibly associated with poor indoor climate and stress.

Occupational skin problems include skin irritations and allergic dermatitis. Occupational dermatitis affects virtually all industries and business. Symptoms include redness, itching, scaling and blistering of the skin and it is very painful. In more severe cases, the skin can crack and bleed and it can spread over the whole body. The affected worker may have to take sick leave or even change or leave their job. The vast majority of cases result from skin exposure to chemicals — dusts, liquids and fumes. Onset following exposure to some substances can be very quick, but it can also take weeks, months or even years. Excessive exposure to just water alone produces skin dryness and chapping, as it removes skin oils, and this in turn can increase the possibility of skin irritation. This effect is even greater where there is exposure to soap and detergents or solvents, so work-related skin diseases are common in 'wet-work' that requires frequent handling of water and cleaning agents. Working where the air humidity is very low also dries the skin (for example, see Meding and Torén, 2001).

Box 28

'My hands were all cuts and blisters. I had to get my boyfriend to help me eat because I could not hold a knife and fork — could not grip anything [...]'.

DM, former apprentice hairdresser.

Source: 'Preventing dermatitis at work', Health and Safety Executive (2002b).

Sectors with the highest risk of work-related skin disease include hairdressing and beauty care; catering and food processing; cleaning; construction; engineering; printing; chemical; healthcare; agriculture and horticulture; rubber and offshore. Some of these sectors employ a particularly high proportion of women. Many of these jobs are so-called 'wet jobs' and women are more often exposed to water, detergents and solvents, for example in jobs such as in hairdressing, dental nursing, cleaning and nursing and caring work. Also, women involved in food handling, farm work and bakery work are at increased risk. Allergic reaction to latex in protective gloves can include skin reactions and more rarely respiratory disorders — rhinitis or asthma (Smith et al., 2002; Wrangsö et al., 2001). Latex allergy has been prevalent among dental and healthcare workers, for example, and NIOSH estimates that, among healthcare workers who experience frequent latex exposure, 8–12 % develop sensitivity to latex (NIOSH, 2001). It is therefore no surprise that women report more occupational skin diseases than men.

> **Box 29**
>
> In Finland, 60 % of cases of occupational skin diseases reported in 2000 were among women (Ammattitaudit, 2000). Female cases of occupational skin diseases are distributed among sectors as follows: healthcare and social care (28.5 %), services (25.5 %), agriculture and forestry (19.9 %) and manufacturing (14.2 %). For male cases, the industries are manufacturing (61.4 %), agriculture and forestry (17.4 %), services (7.0 %), and social care (3.3 %). Chemical exposures account for 86 % of women's skin diseases while for men the equivalent percentage is 91 %. Detergents, the third biggest cause of skin diseases for women, is in ninth place for men while men's third biggest cause of skin diseases is organic solvents, which is ninth for women.
>
> *Source:* Estlander and Jolanki (2003).

There are simple precautions that can be taken to avoid the use of harmful substances where possible, for example by using less harmful ones, safe working practices, the use of good hand washing and suitable protective gloves and other personal protection, good hand care including the use of moisturising creams and carrying out skin checks for early detection. Particularly in some of the small businesses where women are employed, such as hairdressing, there may be a lack of awareness, resources and motivation to take the preventive steps needed.

While the occupational skin problems discussed above are recognised and well documented, skin problems are also among the symptoms of sick-building syndrome, and as discussed in the section on respiratory disorders, sick-building syndrome is more prevalent among women. VDU work is also occasionally associated with skin problems such as itching and rashes — possibly associated with poor indoor climate and stress. The symptoms are mainly reported in the Nordic countries. Electrostatic or magnetic fields from VDUs are not causes, according to current research knowledge (Grönkvist and Lagerlöf, 1999).

To conclude, the higher incidence of dermatitis among women reflects the jobs they do and therefore the occupational exposure to substances that cause skin irritation and allergies.

Work-related cancer

Key points

- There are links between certain cancers and certain occupations.
- Generally, occupational cancer is more common among men than women, in particular due to differences in length and type of exposure.
- Occupational cancer among women is probably underestimated, for example due to poorer record keeping on women's occupations on death certificates.
- Some cancers are more prevalent in occupations dominated by women.
- There are gaps in knowledge, as many occupational cancer studies have not taken account of gender or female-dominated occupations.
- Improved data collection is needed to improve assessment of work exposure to carcinogens among women and therefore prevention.

Incidence and types of occupational cancers

There has been a considerable amount of research into work-related cancer and it is estimated that it accounts for 2–5 % of all cancer cases. Generally, occupational cancer is more common among men than women, in particular due to differences in length and type of exposure, but lack of information may lead to underestimation of incidence among women and this will be discussed later in this section.

A number of chemicals, industrial processes and occupations have been causally linked with cancer in humans. The body parts affected include the epithelial lining of the respiratory organs (nasal cavity, paranasal sinuses, larynx and lung), the urinary tract (renal parenchyma, renal pelvis and bladder), the mesothelial linings, bone marrow, liver and reproductive system.

It has been estimated that in the Nordic countries alone, 3.7 million men and 0.2 million women were potentially exposed to above-average levels of one or more industrial carcinogens in the time period 1970–84. It was expected that these exposures would result in about 1 890 new cases of cancer among men (3 % of all cancers) and less than 25 among women (0.1 % of all cancers) every year from 2000 onwards. The fractions attributable to occupational exposure were estimated to be the following: 70 % of mesotheliomas, 20 % of cancers of the nasal cavity and sinuses, 12 % of lung cancers, 5 % of laryngeal cancers, 2 % of urinary bladder cancers, 1 % of leukaemias, and 1 % of renal cancers (Dreyer et al., 1997).

Links between certain cancers and certain occupations

Exposure to carcinogens varies according to the industry and occupation, for example the links between elevated incidence of testicular cancer and prostate cancer and exposures in some male-dominated industries are relatively well studied. Certain occupational carcinogens are also found in certain female-dominated industries and occupations and there have been a number of studies and reviews that looked at the incidence of occupational cancer among women. These include the reviews by Zahm and colleagues at the US National Cancer Institute and a major analysis of cancer registers covering 20 years of data from Finland, Sweden, Denmark and Norway by Andersen et al. (1999).

Women working in the health sector may be exposed to antineoplastic drugs, anaesthetic gases, ionising radiation and electromagnetic fields. Cosmetic industry workers and hairdressers may be exposed to formaldehyde, hair dyes and hair sprays, and dry cleaners to tetrachloroethylene and trichloroethylene. In addition, workers in various manufacturing industries may be exposed to certain solvents, asbestos, dyes, fumes and so on, and those working in drug manufacturing may be exposed to certain pharmaceutical substances (for example, see Zahm, 2000; Blair, 1998) In northern Europe, where the majority of pub and restaurant workers are women, exposure to environmental tobacco smoke is a problem. According to a recent Finnish study (Autio et al., 2001), workers in pubs and restaurants are exposed to environmental tobacco smoke for at least four hours

---

Box 30: Male and female occupational exposure to carcinogens

Many of the 150 chemical or biological agents classified as carcinogens are encountered in occupational settings. The risk of developing cancer is influenced by the dose received, the potency of the carcinogen, the presence of other exposures (notably tobacco smoking), and individual susceptibility. Globally, about 20–30 % of the male and 5–20 % of the female working-age population (people aged 15–64 years) may have been exposed during their working lives to lung carcinogens, including asbestos, arsenic, beryllium, cadmium, chromium, diesel exhaust, nickel and silica. Worldwide, these occupational exposures account for about 10.3 % of cancer of the lung, trachea and bronchus, which is the most frequent occupational cancer. About 2.4 % of leukaemia is attributable to occupational exposures worldwide.

Source: WHO (2002b), *World Health Report 2002.*

daily, and the proportion of smokers among them is 15–20 % higher when compared to the rest of the population. This means that lung cancer rates are elevated in this group. An examination of the Nordic cancer register data (Andersen et al., 1999) found that women working in restaurants and pubs and in the tobacco and textile industries had higher cancer rates than women working in other industries. Other occupations with elevated risks of cancer included food service workers (lung cancer), and workers producing beverages.

Certain cancers, such as leukaemia and cancers of the lung and bladder, appear to have a strong occupational component. The link between mesotheliomas and asbestos is well known. Leukaemia has been observed to be more common among women exposed to benzene, vinyl chloride, antineoplastic drugs, and radiation. Nasal cancers may result from an exposure to nickel compounds and hardwood dust. Elevated lung cancer rates are associated with exposure to asbestos, certain metals (for example, nickel and chromium-6), and environmental tobacco smoke, and lung cancer among women has been linked to the furniture, asbestos and food industries (Blair, 1998). Textile workers, dry cleaners, and rubber and plastic workers may be at increased risk of contracting bladder cancer.

Smoking is a causal or contributory factor in many incidents of lung cancer among women. Increased smoking is also linked to stressful work, where smoking may be used as a coping strategy. The stressful nature of low-control and monotonous work characterising many women's jobs is discussed in the section on stress.

Box 31: Selected industries, known or potential carcinogens and reported cancer excesses

Industry	Known/potential carcinogens	Cancers
• Agriculture	Pesticides, sunlight, fuels	Brain, cervix, gall bladder, leukaemia, liver, lymphoma, multiple myeloma, ovary, stomach
Service industries		
• Hair and beauty	Hair dyes, hair sprays, formaldehyde	Bladder, brain, leukaemia, lung, lymphoma, ovary
• Dry cleaning	Carbon tetrachloride, trichloroethylene, tetrachloroethylene, other solvents	Bladder, cervix, oesophagus, kidney, liver, lung, ovary, pancreas
• Food service	Tobacco smoke, cooking fumes	Bladder, cervix, oesophagus, lung
• Healthcare	Antineoplastic drugs, anaesthetic gases, ionising radiation, viruses	Bladder, brain, breast, lung, lymphoma
Manufacturing		
• Computers, electronics	Solvents, metal fumes	Brain
• Furniture	Wood dust, solvents, glues, formaldehyde	Lung, pancreas, sinonasal
• Motor vehicle manufacturing	Paints, fumes, solvents, machine fluids	Colorectal, lung, stomach
• Chemical/ plastics/ rubber	Vinyl chloride, 1,3-butadiene, benzene, other solvents, nitrosamines	Bladder, brain, breast, leukaemia, lung, lymphoma, ovary
• Textile apparel	Asbestos, dyes, lubricating oils	Biliary tract, bladder, leukaemia, lung, lymphoma, mesothelioma

Source: Zahm et al. (2000).

Elevated rates of breast cancer have been found among professional women pursuing careers, such as teachers (Ruben et al., 1993), possibly related to hormonal influences attributed to delayed childbearing. In addition, studies have shown an increased risk of breast cancer among women working with pesticides, solvents and in healthcare. There is also evidence that flight attendants may be at increased risk (Zahm, 2000). Exposure to artificial light at night has been hypothesised to increase the risk of breast cancer, as the nocturnal production of melatonin is disturbed (Davis et al., 2001; Schernhammer et al., 2001; Hansen, 2001).

In all, not enough is known about the causes of breast and cervical cancers and, in particular, occupational factors linked to their development. NIOSH is therefore conducting studies on women exposed to a number of substances at work to determine whether there is a link to cancers that affect women, such as cervical and breast cancer. These include ethylene oxide, used to sterilise medical supplies, breast cancer in relation to exposure to polychlorinated biphenyl compounds (PCBs) and cervical cancer in relation to exposure to the solvent perchloroethylene (PERC) used in the dry-cleaning industry (NIOSH, 2001).

Box 31 presents some female-dominated industries, carcinogens found in these workplaces and elevated rates of cancer found among the women workers.

Information on work-related cancer among women

As stated at the beginning of this section, although occupational cancer is more common among men than women, lack of information may be leading to an underestimation of the incidence of occupational cancer among women. Studies on men are far more numerous than studies on women for this topic (for example, see Zahm et al., 1994). Zahm et al. (1994, 2000) and others such as Messing (1998) report various reasons for the imbalance in studies, for example:

- Assumptions that occupational cancers are more prevalent in men, so women are excluded from studies.
- Similarly, many studies have focused on male-dominated jobs.
- Lack of data and other methodological issues. In general, there are many difficulties in investigating occupational cancer, related to multi-factorial causes and exposures and the difficulties in investigating diseases that develop over a long time period. However, there are some specific issues relating to research on women. For example, many death records of women do not contain information about their profession; lack of adequate disease registries; difficulty tracing women over time; and lack of appropriate comparison population disease rates for working women.
- Change in exposure of women to occupational carcinogens — much information on occupational cancers comes from data gathered at a time when the number of women exposed to carcinogens at work was far lower, for example before the growth of the semiconductor industry.

It seems that gender is still being ignored in investigations into possible health effects of new technologies, products and processes entering daily and working lives. The possible health effects of using cellular phones, including the possibility of cancer linked to their electromagnetic fields, are one example where so far the majority of the studies have been gender neutral (see Box 32).

Improving data collection and research on occupational cancer among women

Occupational cancers among women and men are preventable. Data collection and research requirements to identify risks to women and to

> **Box 32: Mobile phones and cancer studies — an example of gender neutrality in new risk investigations**
>
> So far, the great majority of epidemiological and laboratory studies have suggested that there is no link between the use of a mobile phone and brain tumours (Boice and McLaughlin, 2002; Auvinen et al., 2002; Johansen et al., 2001). However, there are contradictory results as well: for example, according to an *in vitro* study conducted by Leszcynski et al. (2002), exposure to mobile phone radiation changes the phosphorylation status of numerous proteins and it may be possible that these changes, when occurring repeatedly over a long period of time, are dangerous. Another epidemiological investigation conducted by Hardell et al. (2002) claims that the use of NMT (Nordic Mobile Telephone, in use until 2000) increases the risk of brain cancer but so far no such conclusions have been made on GSM (global system for mobile communications). Because of several shortcomings in the study design, however, the role of bias and chance may explain the reported findings. Mobile phone studies have ignored gender, apart from a few exceptions (for example, Hietanen et al. (2002), a study using a small sample which looked at the occurrence and nature of symptoms reported by 20 people — 13 women and 7 men — claiming to be hypersensitive to electromagnetic fields). The large number of confounding variables, short research history and lack of tests covering large populations can partly explain this. However, there may also be assumptions that there will be no gender differences, so gender is ignored in the studies.

improve assessment of workplace exposure include:

- inclusion of occupation and gender in routine data collection such as cancer registries and computerised death certificate data;
- further research on jobs dominated by women, for example manufacturing with potentially harmful exposures, including in the semiconductor industry;
- better study designs and research tools to deal with some of the methodological issues such as small sample sizes of exposed women;
- analyses based on specific task and type of exposure, not just by job title or sector;
- analyses of data for women in existing studies where data have been collected but never analysed.

Hearing disorders

Key points

- Male workers suffer more noise-induced hearing problems than women workers. Men are more exposed to high levels of noise at work than women, due to job segregation.
- Female-dominated jobs associated with high noise exposure include those in food production, bottling and textile manufacturing. Non-traditional noisy workplaces for women and men are where loud music is played such as in clubs and discos.
- Other workplace noise issues, such as stress from noise below levels that can produce hearing damage, and oversensitive hearing warrant further investigation.
- Most studies have been based on male populations, and physicians may fail to link cases of hearing loss in women to work.
- There is some evidence of an emerging risk, acoustic shock, to which women may be more exposed, which causes different symptoms and has an effect at lower exposure levels than 'traditional' noise-induced hearing loss problems.

Noise-induced hearing disorders and their prevalence

Noise is a 'traditional' hazard in the industrialised workplace. Too much noise exposure may cause a temporary change in hearing (a feeling of not being able to hear properly or blocked ears) or temporary tinnitus (ringing, whistling or buzzing in the ears). These temporary problems from short exposures tend to go away within a few hours of leaving the source of the noise.

However, repeated exposures to loud noise can lead to permanent, incurable hearing loss, tinnitus, or oversensitivity to noise. Permanent hearing loss can also be caused by immediate exposure to an extremely loud noise such as an explosion. Noise in the workplace can also cause stress and be a safety hazard if it interferes with communication. Noise-induced hearing loss develops gradually. The results are that the person can have difficulty in following conversations in a group, answering the telephone or his or her family complains that the volume of the television is too high. There are also occupations where the employees are exposed to disturbing noise every day, although it is not at a level to damage hearing, for example schoolteachers. This kind of environmental noise can be disturbing, be an additional source of stress and cause, for example, sleeping problems, lack of concentration, depression and aggressiveness.

The development of noise-induced hearing loss is related to characteristics of the noise such as the level of noise, duration of daily exposure, type of noise (continuous or impact noise) (Campo and Lataye, 1992), exposure to toxic drugs and certain solvents (Morata et al. 1991; Myers and Bernstein, 1965) and individual susceptibility. Biological and human-related factors (Humes, 1984; Pyykkö et al., 1986; Borg et al., 1992) and genetic factors (Barrenäs, 1998; Gates et al., 1999; Kaksonen et al., 1998) have also been reported, including age and gender.

Work-related hearing disorders are more prevalent among men than women. In the European labour force survey, men reported more than women in all Member States. In the EU as a whole, it was reported as the most serious work-related complaint by 4.2 % of men and 1 % of women (Dupré, 2002). Analysis of the third European survey on working conditions by full and part-time work showed a drop from 11 % of full-time male workers reporting suffering a work-related hearing problem to 6 % for men working part-time. The figures are 3 % for full-time female workers and 2 % for part-timers (Fagan and Burchell, 2002).

This gender difference is due to job segregation. Men are more likely to work in environments where noisy plant, equipment or tools are used, such as construction, road maintenance, mining, saw mills, foundries, and engineering workshops. Many authors have found a significant and relatively large difference in the incidence of hearing disorders between men and women (Berger et al., 1978; ISO, 1990). However, once differences in noise exposure are controlled for, it appears that no difference is found (Davis et al., 1998).

While noise-induced hearing disorders are less common among women, there are some female-dominated industries where noise is a problem: for example, food processing was the industry with the largest number of women with reported hearing disorders in Finland in 2000. Altogether, 801 men and 54 women were reported to have noise-induced hearing loss (Ammattitaudit, 2001). Bottling factories and textile manufacturing are particularly noisy environments where women may work. Both women and men may work in environments where loud music is played.

An acoustic shock incident is defined as a sudden increase in high-frequency noise transmitted through a headset. In call centre work, for example, employees can be exposed in relation to interference on the telephone line. Although call handlers may be shocked or startled by the noise, exposure to these unexpected acoustic events should not cause hearing damage as assessed by conventional methods. However, emerging evidence suggests that exposure to these acoustic incidents, at levels much lower than is traditionally associated with hearing damage, is giving rise to other symptoms (HELA, 2001; Risking Acoustic Shock Conference, 2001; see also Box 35).

> **Box 33: Tinnitus and gender**
>
> There is some evidence that there may be gender differences in the experience of tinnitus. One study found that the description of tinnitus sounds varied between the two sexes: the description most used by men was ringing. Women described the sound more often as chirping. Furthermore, women's tinnitus was more concentrated on one ear only, and more tension was diagnosed in the cervical area. After treating the patients with injections in the muscle triggering the tinnitus, it was concluded that the outcome of the treatment correlated positively with female gender and the type of tinnitus (chirping) they had.
>
> *Source:* Estola-Partanen (2000).

Noise and reproductive hazards

The possible effect of noise on the foetus is discussed in the section on reproductive health.

Prevention and detection

The measures needed to prevent and control exposure to workplace noise are well known. In general, the number of workers suffering from noise-induced hearing loss has decreased due to reductions in some areas of heavy industry and also better use of preventive measures. These include the use of quieter machinery, separating the noisy equipment from workers, and, as a last resort, the use of hearing protectors.

There may be an issue for women if earmuffs designed for the standard (male) head are a bad fit for people with smaller heads. Training in the correct use of earmuffs needs to take account of the correct fitting for people with long hair or jewellery. Risk assessments and inclusion in hearing surveillance programmes should be based on looking at real exposure and not making assumptions based on job titles.

While this review did not find information to support any differences in sensitivity to levels of occupational noise, it should be noted that much of the research contributing to current standards on exposure to noise will have been based on male-dominated studies. There may be some underestimation of work-related hearing among women, particularly where there is also an ageing element in its onset and physicians do not enquire about possible work exposures.

> **Box 34**
>
> A Portuguese national campaign to improve safety in the textile sector included prevention of risks related to noise as one of the priorities. Actions included prevention manuals for specific activities, posters, leaflets, etc., and bulletins covering technical solutions for specific risks.
>
> *Source:* European Agency, 2001a.

> **Box 35: Improving the acoustic safety and comfort in call centres**
>
> - According to a study of operators, the main sources of background noise, within the call centres, were other call handlers speaking to customers, staff talking to each other, particularly at shift changes, meetings being held and telephones ringing in other parts of the open-plan office. At the caller's end, the television, radio, dogs barking and babies crying were cited. Call handlers' overall daily personal noise exposure is unlikely to exceed 85 dB(A). However, there are still a variety of measures that can be taken to reduce their exposure to noise at work and improve the acoustic comfort of their work:
>
> - Train call handlers to place the microphone in the optimal position in front of a call handler's mouth to avoid excessive vocal feedback for both the caller and the call handler. Callers may become frustrated if they cannot hear call handlers clearly, and there is a risk that call handlers may start to strain their voices in order to be heard.
>
> - Special noise absorbing material, often in ceilings, can help to reduce reverberation in the call centre.

- Carpet, chairs with soft seats and padded screens between call handlers can also be effective noise absorbers if designed and fitted appropriately.
- To limit call handlers' daily personal noise exposure, headsets, amplifiers and/or turrets should be fitted with volume controls, and call handlers should be trained in how to use the volume controls. There is a risk that call handlers will turn the volume up in order to hear a quiet caller but forget to turn it down for the next caller even if that caller speaks at a higher level. Call handlers may then get used to listening to callers at higher levels than are really necessary. Some systems return the call handler's listening level to a default setting after each call. An on-screen reminder at the start of a new call could also prompt call handlers to assess the level and adjust the volume if necessary. A designated key on the keyboard can be used to reduce headset noise immediately to a minimum when pressed. This is a very quick method for call handlers to reduce sudden high headset noise levels. Prompt call handlers to adjust the listening level (both up and down) through their headsets at the beginning of each call.
- Call handlers should be encouraged to report to management exposure to all acoustic shock incidents or any other abnormally loud noises. Employers should make a record of these reported events.
- One practical way of limiting exposure to unexpected high noises from headsets is through headset design, such as the incorporation of an acoustic (shock) limiter. A limiter ensures that any type of noise (for example, conversation, short duration impulses) above a fixed volume is not transmitted through the headset.
- Call handlers wear a headset throughout their shift every shift so it is important that it is fully adjustable to ensure a comfortable fit. Headsets should be checked regularly and repaired or replaced immediately if necessary. There may be an increased risk of ear irritation and infection because headsets are worn so intensively. To reduce this risk, staff should be trained in headset hygiene and given the time and the materials to complete a hygiene programme. The issue of headsets to individuals is strongly recommended.
- Voice tubes can become blocked with food, make-up and dust, and this compromises the effectiveness of microphones. Call handlers must be trained in how to clean the voice tubes in order to optimise the volume of the transmitted signals and avoid the risk of frustrated callers and strained voices.
- Ensure a sufficient stock of new or sterile headset pads and voice tubes is maintained.
- Employees should be encouraged to report immediately exposure to any acoustic incident that results in physical damage. Employers should implement a policy so that the details of these incidents are recorded, and employees are examined by an appropriate expert to investigate the extent of any physical damage (this may include a hearing check).
- Call handlers should be provided with information about the potential risks to hearing and the measures being taken by their employer to control these risks. Call handlers or their representatives should be consulted about working practices.

Source: HELA (2001).

Vibration-related diseases

Key points

- Male workers suffer more vibration-related diseases. Men are much more exposed to vibration at work than women, due to job segregation.
- Further study of prognostic factors among women is needed as well as more study to investigate whether risk varies between women and men with the same occupational exposures.

Vibration exposure occurs in many occupations, from contact with vibrating machinery or equipment. Hand–arm vibration causes damage to blood vessels and nerves in the fingers. The resulting condition is known as white finger disease, Raynaud's phenomenon or hand–arm vibration syndrome (HAVS). The affected fingers turn white, especially when exposed to cold. Vibration-induced white finger disease also causes a loss of grip force and reduced sensitivity to touch. The effect of whole-body vibration is poorly understood. Studies of drivers of heavy vehicles have revealed an increased incidence of disorders of the bowel and the circulatory, musculoskeletal and neurological systems.

According to data from the third European survey on working conditions, men are much more exposed to vibrations at work than women. Nevertheless, 7 % of female respondents reported being exposed at least half of their time at work to vibrations from hand tools, machinery, etc. The figure for male respondents was 23 %.

Sewing machines and looms in the textile sector are one source of exposure to vibrations in female-dominated jobs. One Swedish study, following up a sample of women who had reported work-related hand–arm vibration injuries to the Social Insurance Office or had received compensation, found the highest prevalence was among dental technicians using handheld vibrating tools (Bylund et al., 2002).

There is also evidence of an increased risk of work-related upper limb disorders in occupational groups working with vibrating tools (Buckle and Devereux, 1999). Bylund and Burström (2003) cite several studies that indicate that women may have a higher rate of symptoms from the use of vibrating tools than men. The issue of appropriate ergonomically designed tools for women users has already been discussed. Where women use vibrating tools that are ergonomically poorly designed, could they be at increased risk of vibration disorders and upper limb disorders compared to men? Anthropometric factors linked to higher rates of vibration injuries include being short and use of higher grip forces (studies cited by Bylund and Burström, 2003). They carried out their own study, that took account not only of vibration level, frequency and duration, according to the international standard for vibration exposure measurement, but also handgrip forces, postures and individual factors. According to their results, once anthropometrical differences were taken into account, no gender differences in power absorption rates were found. Research on vibration injuries has predominantly been carried out on men. More research is needed to investigate further both the ergonomic issues and gender differences in susceptibility. There may be risks of menstrual disturbances and spontaneous abortions among women exposed to whole-body vibrations and this should be investigated further.

Working temperatures

Key points

- Men report more exposure to hot and cold working conditions than women.
- It is important to look at the nature of the work being done in hot and cold conditions.

According to the third European survey on working conditions, men are more likely to report being exposed to either hot or cold temperatures for at least half their time at work than women; 11 % of women respondents reported working in high temperatures and 7 % in low temperatures. The corresponding figures for men were 17 and 16 %.

Work areas where women may be exposed to hot temperatures include kitchens and laundries and agriculture. As well as agriculture, food processing is one area where women may be exposed to cold temperatures.

Women are probably very under-represented in the extremes of exposure to noise, damp, cold or heat at work. However, for example, a study in the poultry-processing industry found that women were not exposed to the extreme cold of freezers, but they were in relatively immobile postures at around four degrees, whereas men walked back and forth from extreme cold to room temperature. Women were more likely to report cold extremities (Messing et al., 1998b; Messing, 1999). It is also important to look at the type of work being done, as, for example, it appears that working in cold temperatures may increase the risk of work-related upper limb disorders.

Reproductive health

Key points

- Most research on reproductive health has concentrated on pregnant women and foetal protection and only pregnant women and new mothers are covered by specific EU legislation.
- The possible influence of working conditions on some women's health issues, such as menstruation and the menopause, has been almost completely ignored.
- More research and workplace guidance are needed on the effect of work on other reproductive hazards relating to both women and men, including fertility, sexuality, early menopause, and menstrual disorders.
- Incorporating indicators of reproductive health complaints into European surveys of working conditions outcomes could be considered.
- Existing good practice for research methodology and risk assessment and prevention should be shared.

Clearly, the reproductive systems of women and men are different and this means that the reproductive health of women and men is affected differently by exposures in the workplace. Reproductive disorders include birth defects, developmental disorders, spontaneous abortion, early foetal death, low birth weight, preterm birth, and congenital defects and illness in the offspring; they also include reduced fertility, impotence, and menstrual disorders. Particularly for women, these issues should be dealt with in the context of their life, because they are related to age (White, 2000).

In contrast to some other occupational safety and health areas, work hazards in relation to reproductive health have been most studied in relation to women. For example, in an analysis of recent journal articles (see Box 57 in Section 5), 93 of 251 articles on women's occupational health dealt only with reproductive health, while for men it was only 15 of 314 articles.

The focus has been very much on pregnancy, looking at the woman as the child bearer and carer, and the health of the foetus and newborn child. There have even been examples of legislative 'foetal protection policies' that exclude women of fertile age from employment. Even if well meaning, these policies ignore the fact that men are also exposed to prenatal reproductive hazards and restrict the view of women's reproductive health (Stellman and Lucas, 2000). Consequently, less attention has been paid to the effect of work on male and female fertility, sexual and reproductive functioning, and issues concerning new and nursing mothers.

While there have been individual studies looking at the reproductive health effects of particular workplace hazards, there seem to be no overall figures about the extent to which work affects reproductive health. The European labour force surveys on working conditions have not included questions in this area. Although numerous occupational exposures have been demonstrated to impair fertility (e.g. lead, solvents and some pesticides), the overall contribution of occupational exposures to male and female infertility is unknown.

Pregnant women and new mothers

Work exposure of pregnant women can affect their offspring in various ways. Chemical substances that enter the mother's bloodstream can pass into breast milk. Poor working conditions can contribute to a range of health issues for women associated with pregnancy and the period after giving birth, including tiredness and fatigue, pain and discomfort, back problems, postural problems, swollen legs, nausea, blood pressure changes and stress.

> **Box 36: Hairdressing and low birth weight babies**
>
> Hairdressers are exposed to many chemicals in their work. This may cause allergies, asthma, hand eczema and other health problems. Furthermore, according to a study of over 7 000 Swedish hairdressers, they also have a slight increased risk of having low birth weight babies, and infants with major malformation compared with women from the general population. The authors found that frequent exposure to chemicals used in permanent waving and hair spray tended to be associated with increased risk of having a low birth weight baby (Rylander et al., 2002). However, there are other factors that could also contribute within the hairdressers' work, such as prolonged static standing.

The EU has a specific directive on the protection of pregnant workers and workers who have recently given birth or are breastfeeding (Council Directive 92/85/EEC). Employers are obliged to carry out a risk assessment for these workers. They should first act to remove hazards and avoid or reduce the risks found. If it is not possible to remove the risks, the directive requires the woman to be offered alternative work, or if that is not possible, she should be suspended from work with maintenance of salary, although there are arguments that the directive should be strengthened concerning obligations to remove risks at source (for example, see Vogel, 2002). The directive also includes measures for maternity leave, time off for antenatal examinations and protection from dismissal. In addition, the EU requires the labelling of chemicals for reproductive toxicity.

The directive recognises a broad range of chemical, physical and biological agents, work processes and working conditions that could present a risk for new and expectant mothers. The European Community has produced additional guidelines on risk assessment to support the implementation of the directive (European Commission, 2000a). A list of the factors that the guidelines cover is given in Box 37. However, the guidelines do not mention issues such as postnatal depression, which can be a serious problem for some women returning to work.

Many Member States have produced additional good practice guidelines and protocols to support their own regulations brought in to comply with the directive. For example, Spain has produced good practice guidelines, a negotiating guide and protocol and guidelines for inspectors (Spanish Ministry of Employment and Social Affairs, 2002). Italian activities have included guidelines on assessing risks related to pregnancy from noise, vibrations and antineoplastic drugs (ISPESL, 2002). It should also be noted that, in workplaces, exposure of all workers to many of these hazards should already be controlled according to the requirements of more general European Union directives, for example covering dangerous substances. Properly implementing the general and specific legislation and guidelines in workplaces would go a long way to improving the health of pregnant women and the foetus. This could be supported by improved awareness of the types of jobs and tasks where there could be risks, as well as studies and the development of practical tools on effective risk assessment and intervention. Existing good practice should be shared. Awareness of rights and benefits available is important too. In Finland, there have been findings that the take-up rate of special leave for pregnant women has been lower than expected due to lack of awareness of employers and employees.

Risks in this area and means of preventing them still need to be studied further, for example adequate testing of chemicals for reproductive effects and the effects of repetitive work, fast-paced work, cramp postures and static effort. More research attention also needs to be given

Box 37: Hazards and issues covered in the European Commission guidelines on the assessment of risks to pregnant workers and workers who have recently given birth or are breastfeeding

- Mental and physical fatigue and working time (long hours, night work and shift work)
- Postural problems connected with the activity of new or expectant mothers
- Working at heights
- Working alone
- Occupational stress
- Standing activities
- Sitting activities
- Lack of rest and other welfare facilities in the workplace
- Risk of infection or kidney disease as a result of inadequate hygiene facilities
- Hazards as a result of inappropriate nutrition
- Hazards as a result of unsuitable or absent facilities related to breastfeeding and expressing milk
- Shocks, vibration or movement
- Ionising radiation
- Non-ionising radiation
- Extremes of cold or heat
- Work in hyperbaric atmosphere
- Biological agents
- Chemical agents, including: mercury, antimitotic (cytotoxic) drugs; substances that can be absorbed through the skin (includes some pesticides); carbon monoxide; lead
- Manual handling of loads
- Movements and postures
- Travelling, inside and outside the workplace
- Underground extractive industries
- Work with display screen equipment (VDUs)
- Work equipment and personal protective equipment and clothing

Source: European Commission (2000a).

Box 38

Risks to the reproductive health of women in the healthcare sector include exposure to:

- waste anaesthetic gases;
- antineoplastic drugs;
- disinfectants and other chemical agents (e.g. mercury among dental workers);
- non-ionising radiation;
- ionising radiation;
- shift work and night work;
- heavy physical work, manual handling and prolonged standing;
- biological agents;
- physical and psychosocial strain.

Sources: Figà-Talamanca (2000); Lindbohm and Taskinen (2000); Taskinen et al. (1999); Sallmén and Taskinen (1998); Axelsson et al. (1996).

to the effects of working conditions on the health of the woman herself, and the effectiveness of legislative and supporting measures should be assessed. The rest of this section will look at those areas of reproductive health that have received rather less attention and are not covered by specific legislation.

Menstrual disorders and fertility

Menstrual disorders include dysmenorrhoea — severe menstrual pain and cramps, amenorrhoea — absence of menstruation, irregular cycles, anovulatory cycles and reduction in fertility. Menstruation is not openly talked about, even among women, and it is highly unlikely to appear among the possible list of work-related health symptoms in checklists and surveys that are commonly used to assist workplace risk assessments. If the question does appear, women may still be reluctant to answer it (Messing,

1998) and, in addition, women are reluctant to draw attention to 'women's problems' in the workplace. Despite the fact that it is relatively easy to carry out studies looking at menstrual health compared to other reproductive health issues, there is a paucity of research on the menstrual cycle in relation to work exposures (Messing, 1998).

Some work has been done and, for example, Figà-Talamanca (1999) cites the following work exposures that have been associated with menstrual disorders:

- strenuous physical work associated with disruptions of the menstrual cycle;
- demanding and stressful work showing menstrual effects in women workers in sectors including agriculture, industry, service sector, nursing, air transport, armed forces, etc.;
- occupational stress associated with dysmenorrhoea;
- exposure to hormones and alkylating agents, for example in the pharmaceutical industry associated with menstrual disorders;
- exposure to halogenated hydrocarbons and organophospates, for example in the production and use of pesticides, associated with menstrual problems (see also Valls-Llobet, 2002);
- exposure to heavy metals such as lead, mercury and cadmium in the metal mechanic industry associated with abnormal menses and amenorrhoea;
- exposure of dental workers to mercury used in dental amalgams associated with reduced ability to conceive;
- exposure to environmental noise and hot and cold working conditions in the food industry associated with dysmenorrhoea, hormonal disturbances and reduced fertility;
- exposure to solvents, such as benzene, styrene, carbon disulphide and formaldehyde associated with menstrual and ovarian function disturbances and reduced fertility;
- exposure to solvents, including 'within acceptable exposure standards' for example styrene, xylene, toluene, trichloroethylene, tetrachloroethylene, which are used, for example, in dry cleaning and laboratories associated with delays in conception.

In addition, NIOSH (1999b) mentions that exposure to carbon disulphide (CS_2) is associated with menstrual cycle changes among viscose rayon workers.

Women commonly experience pre-menstrual symptoms. Pre-menstrual syndrome consists of many symptoms such as mood swings, forgetfulness, lack of concentration, tiredness, appetite changes, abdominal bloating, sore breasts, swelling of hands and feet, hot flushes, muscle aches, headaches and migraine. Exposure to workplace stressors may contribute to or increase the symptoms of pre-menstrual syndrome as stress affects the menstrual cycle

Box 39: An example of a gender-sensitive approach to fertility studies

Sallmén (2000) investigated whether occupational exposure to organic solvents or inorganic lead is associated with reduced fertility. Data on time to pregnancy were collected both from female workers and wives of male workers. Daily or high maternal exposure to solvents correlated with reduced fertility. In particular, halogenated hydrocarbons and aliphatic hydrocarbons increased the risk, and prolonged times to pregnancies were observed among women working in dry-cleaners, shoe factories and the metal industry. High maternal exposure to lead was only tentatively associated with prolonged time to pregnancy. The study also provides limited support for the hypothesis that paternal exposure to organic solvents reduces fertility and that exposure to inorganic lead is associated with delayed conception. The potential biasing effect of restriction to couples with pregnancies was investigated by including a register-based study on the occurrence of marital pregnancies among the wives of the monitored men. The findings of the study strengthened the conclusion.

(Hatch et al., 1999). Reducing stress may relieve symptoms.

One way of studying fertility is to measure fecundability — i.e. the probability of clinical pregnancy in a given menstrual cycle. However, see Baird and Strassmann (2000) for information about the limitations of some existing studies on fertility.

Endometriosis is a hormonal and immune condition in which the cells lining the interior of the uterus (endometrium) are found in other parts of the body. It produces a variety of symptoms including menstrual disorders, moderate or severe abdominal pain, bowel problems, or infertility. The cause is not known, but any investigations into environmental causes should take account of work exposures.

Libido (sexual drive) and working hours

Most research into reproductive health has concerned physical, chemical and biological agents. Limited attention has been given to the effect of psychosocial risk exposures on reproductive health. One study has looked at the relationship between working hours and libido or sexual drive (Pryce et al., submitted). A relationship was found between long working hours and reported impairments to libido for men, and between inflexible working hours and impaired libido for women. However, as the researchers point out, the study did not take account of the unpaid work in the home, and this should be taken into consideration in future research. In a survey of the health of women workers in the Tuscany region in Italy, there were reports of reduction in libido, as well as menstrual disorders, in relation to workplace stress (Massai, 2001).

Menopause

Little attention has been paid to the effect of work exposures in relation to early onset of the menopause. However, some studies suggest associations between age at natural menopause and environmental exposures. For example, Wise et al. (2002) found a link between economic hardship and early onset of the menopause and they suggested that it was probably a combination of factors including stress and exposure to tobacco smoke and other toxins. Cooper et al. (2002) looked at exposure to plasma polychlorinated biphenyls (PCBs) and 1,1-dichloro-2,2-bis(-chlorophenyl) ethylene (DDE) and age at natural menopause. They found an association between DDE and early age at natural menopause similar to the association between smoking and menopause, but no association was seen with PCBs. They conclude that prospective studies, in which exposure measurements are taken before menopause, would be particularly useful for further research.

In contrast, much more attention has been focused on the possibility of hormonal changes, for example seen at menopause, being the cause of musculoskeletal disorders, such as carpal tunnel syndrome, in women. However, the role of female hormones in the development of musculoskeletal disorders is very poorly understood and the literature is inconsistent. The data do not confirm that hormonal status is an important modifier of the effect of ergonomic exposures at work on musculoskeletal disorders. In particular, it should also be remembered that older women workers are likely to have had longer exposure to the poor ergonomic working conditions that are known to give rise to musculoskeletal disorders (Punnett and Herbert, 2000; Messing, 1998; Doyal, 2002).

One study concerning the menopause and work has been a survey of worker safety representatives in the UK about the issue (Paul, 2003). A question about the menopause was included in a general survey of trade union safety representatives about working conditions and 22 % of respondents said that women at work were raising the menopause as a problem made worse by work. As a result, a further survey to look specifically at the workplace health and safety implications was carried out. The report of the survey mentions various factors associated with the menopause that can be made worse by work or more difficult to cope with, including: hot flushes and sweats; headaches, tiredness and fatigue; stress, anxiety and depression; disturbed sleep patterns; heavy or more unpredictable menstrual bleeding; skin and eye dryness and irritation; side effects of hormone replacement therapy such as nausea and vomiting; urgency to urinate; increased risk of osteoporosis from sedentary work, restricted posture, dietary factors or lack of light (diet and light linked to calcium deficiency).

The survey results suggest a number of ways in which occupational safety and health policies and health-promotion policies can be supportive to women going through the menopause. Measures for prevention and control in the workplace include:

- adjustable working temperature, adjusting humidity and additional, individual ventilation;
- access to suitable rest and toilet facilities and cold drinking water;
- adequate and flexible rest breaks and toilet breaks;
- suitable uniforms (layered and loose clothing) and protective clothing;
- access to natural light;
- providing better seating;
- reducing long hours and stress;
- healthy eating options, for example in canteens or snack vending machines;
- working policies for time off for treatment, flexible working conditions and constructive handling of sickness absence;
- facilitating access to information and advice;
- providing a sympathetic and supportive working environment.

Source: Paul (2003).

Box 40

The ISPESL (Italian National Institute of Occupational Safety and Prevention), in collaboration with various universities and health institutes, launched a project covering all areas of reproductive risks to women workers, including the role of work in relation to the menopause, taking account of the increase in older women at work, and also the influence of work-related stress. The project is also including awareness raising, information and education about work-related reproductive risks as a major element, as such risks are often underestimated. Funded by the Ministry for Health, the project aims are to:

- evaluate reproductive risks among female agricultural workers;
- verify a possible connection between work exposure to antineoplastic drugs and reproductive health, with particular attention to reduction of fertility;
- evaluate the effects of work on age of onset of the menopause, on endocrine regulation, on bone fractures and on the cardiovascular system;
- promote and implement initiatives to increase awareness and education about work-related reproductive health and to improve health services for women in this area;
- conduct research into morbidity and mortality among healthcare workers and their offspring;
- evaluate mental and physical fatigue in women working in healthcare professions in order to improve training and interventions.

Source: Papaleo et al. (2003).

Male reproductive health

The focus on the foetus of the pregnant woman has meant that much less attention has been paid to fertility and genetic damage of both male and female workers. However, organisations such as NIOSH point out that observed global trends in men's decreasing sperm counts have increased concerns about the role of chemicals encountered at work and in the environment at large (NIOSH web site (a). See also

European Commission, 1999). Pesticides, radiation, lead and heat exposure are among the conditions that affect sperm and male fertility (Messing, 1998; NIOSH, 1997). More studies are also needed on the effects of work exposures on reproductive health. However, again, implementation in the workplace of existing general legislation on dangerous substances and physical agents would go a long way to protecting male reproductive health.

> **Box 41: Legislation on reproductive health**
>
> While all EU Member States have introduced legislation on the protection of pregnant women at work, Finland is one country that has introduced occupational safety legislation that covers reproductive hazards more broadly. Risk assessment must take account of 'the potential risks to reproductive health' and exposure to hazardous chemicals, physical agents and biological agents shall be reduced so that 'no risk is caused to the employees' safety or health or reproductive health'. Risk assessment must also take account of 'the employees' age, gender, occupational skills and other personal capacities'.
>
> *Source:* Occupational Safety and Health Act 738/2002: Finnish Ministry of Social Affairs.

The effects of working hours and inflexibility, and responsibility for household duties

Key points

- Work–life balance is important for all working people. Long hours, 'non-standard' elements in work schedules and lack of worker control or flexibility in working hours contribute to the mismatch between work and other aspects of life.
- Policies and practices to support those with caring responsibilities would particularly benefit women.
- Men working full-time also want to work fewer hours and those with families want to take advantage of 'family-friendly' working practices. 'Work–family' policy debates should also address men's work schedules, which should also help women to work in male-dominated areas.
- The legislative and social policy context greatly affects the extent, operation and outcome of 'family-friendly' policies.
- Women's working life is physically and mentally more demanding where they perform a 'double shift' of paid work and unpaid work in the home. A more equal division of responsibilities at home would certainly benefit women, but could bring positive changes for men as well.
- Further investigation is needed into what women and men consider to be 'compatible' schedules and effective policy and practice to improve work–life balance, and examples of good practice should be shared.

In the EU, there has been a growth of households where both partners are in work, and 'dual partnership' households are now more prevalent than 'male only working' households in most Member States (Franco and Winqvist, 2002). As has already been discussed, there is an uneven sharing of unpaid work in the home, with women generally still doing far more, even when both partners work full-time. This means that, for many working women, their everyday working life is both physically and mentally more demanding and stressful as they are carrying out a 'double shift' of paid and unpaid work. Even when ill, women still often have to take care of the needs of their family members and thus may not be able to get sufficient rest to recover.

Various studies have pointed out that this double workload could have serious health and safety consequences, and women are most exposed to this situation (Bielenski and Hartmann, 2000; Fagan and Burchell, 2002; Yeandle et al., 1999; Bercusson and Weiler, 1999; Kauppinen and Kandolin, 1998; Rohlfs et al., 2002; Valls-Llobet et al., 1999; Artazcoz, 2001). Stress and chronic fatigue can be the outcome of the 'double shift' and documented health effects include raised blood pressure (Brisson et al., 1999). Long and inflexible working hours and arrangements are recognised as contributing to work-related stress, causing conflict between work and home life. Lifting, awkward postures and repetitive movements are a feature of many household tasks, although there has been little study of this (Punnett and Herbert, 2000). Cleaners are among those who may be exposed to the same chemical agents at work and in the home.

As already discussed, gender differences in working hours vary between the Member States, including the extent of part-time working and the number of part-time hours worked, and this partly reflects variation between tax systems, childcare provisions and institutional arrangements. For women, employment rates and the number of hours worked vary by age. This will be influenced to a large extent by their greater domestic responsibilities. Research and surveys, for example by the European Foundation, have found evidence of a clear desire and motivation on the part of both men and women to take part in paid employment. However, it was also found that a large proportion of women and men in full-time jobs would like to work shorter hours and devote more time to their private life. Men tend to work the longest working weeks (more than 45 hours) and their ideal would be 37. 'Work–family' incompatibility increases as full-time hours lengthen and is most acute for those working at or in excess of the 48-hour maximum set in the EU working time directive. Preference for part-time work is widespread, particularly among women, although many working part-time would prefer to work more hours. In summary, a large proportion of both men and women workers would like to alter their working hours, to change from either very short part-time work or very long hours to long part-time or moderate full-time hours (see Atkinson, 2000; European Foundation, 2002c; Fagan et al., 2001; Fagan and Burchell, 2002).

The European survey on working conditions found that both men and women were more likely to report that their work schedules are compatible with life outside work if they have 'standard' work schedules of daytime, weekday and fixed hours, as well as not working long full-time hours. Men more often report this incompatibility than women, often because they have 'non-standard' elements in their work schedule in combination with a longer working week (Fagan and Burchell, 2002).

Box 42

Within the labour market, having children strongly influences the participation of women in particular. Available statistics from 11 Member States show that 72 % of childless women aged 20–50 are employed compared to 94 % of men in the same group. This figure drops to 54 % for women with children under six compared to a much smaller drop to 89 % for men in the same group (European Commission, *Social Agenda* 2, 2002).

Box 43

Family-friendly policies are [...] policies that facilitate the reconciliation of work and family life by fostering adequacy of family resources and child development, that facilitate parental choice about work and care, and promote gender equality in employment opportunities (OECD, 2002).

Elements of family-friendly work arrangements can include:

- childcare arrangements such as workplace facilities or financial incentives;
- maternity and paternity leave;
- part-time work;
- leave when children are ill;
- flexible working hours and working arrangements;
- working from home.

There is clearly often a mismatch between work organisation and goals at work on the one hand, and commitments and aims outside of work on the other, and the consequences are often greater for women, but there is also scope for change. Attention should be paid to achieving good work–life balance for all working people. Among measures to support this are policies and practices to support parents and carers, and there appears to be strong support from employees, both female and male, for improvements in this area (for example, ACTU, 2003).

An OECD report points out a number of effects of failure to reconcile work and family life. Parents — or would-be parents — may decide to delay having children, have fewer children than they really want or not have them at all. While some may prefer to drop out of the labour market to care for children, others would like to work or work longer hours but cannot due to inappropriate working hours or lack of childcare, etc. Others may work longer than they want, putting the family under strain, which can even result in broken relationships or young people going without the care they need (OECD, 2002). Family-friendly policies are therefore a benefit to employers, for example by helping them to keep skilled and experienced staff, attract new staff using their positive image and reducing stress on employees with children.

There are also a number of important, broader issues that should be taken into account, but which are outside the scope of this report, such as access to formal and informal childcare, family income in and out of work, as well as family-friendly work arrangements and child-related leave programmes, and there are wide variations in these areas between Member States. For examples and discussion relating to wider issues such as governmental policy, legislation and taxation systems, see OECD (2002). Its detailed report of policies in the Netherlands, Denmark and Australia points out that there can be problems with 'family-friendly' policies, such as increasing gender segregation if they are available only in organisations or sectors that have a female-dominated workforce (see also Gonäs, 2002), and that the legislative and social policy context greatly affects the extent, operation and outcome of 'family-friendly' policies.

In Finland, legislation now allows organisations to arrange childcare services for those employees who would otherwise stay at home to look after a sick child. According to the law, childcare service paid for by the employer is a tax-free

Box 44

A working group at a healthcare trust (hospital and related services), in the UK, developed an approach to working-time and employment practices that enables staff to combine work more easily with the rest of their lives. Once the scheme was developed, it went through the joint staff committee, a forum of managers and staff representatives, before it was finally approved. Scope for flexibility includes part-time working, temporarily reduced working hours, job-share, staggered working hours, annual hours, phased return to work, working from home, career break, special and parental leave and personalised annual leave — an increase in leave entitlement up to a maximum of 40 days a year, with a commensurate reduction in salary or a reduction in entitlement by up to five days for additional salary.

A large supermarket chain in the UK is undertaking a work–life review, which will bring together and standardise a range of strategies that are already in place into a formalised work–life policy. Current benefits include enhanced maternity pay, paternity and parental leave, career breaks, adoption and fostering leave, leave for fertility treatment, time off for emergencies, paid bereavement leave, nursery facilities (in London only), part-time, flexible working and job-share.

Source: *Creating a work–life balance: A good practice guide for employers*, UK Department of Trade and Industry, 2000.

benefit for the employee. However, the benefit does not necessarily concern all employees, but might be granted only for people in key or senior positions in the company. The employee is entitled to two to four days of childcare services, which is the same length of time that an employee would be allowed to stay at home to care for a sick child (Cancedda, 2001).

Where implementing work–life balance practices is on a voluntary basis, it seems that many organisations are yet to be convinced of the benefits — for example, a recent UK study found that only 3 % of the establishments that it surveyed offered any daycare programme for employees' children and only 8 % offered any financial assistance for this (Taylor, 2002). However, there are a growing number of examples of good practice (see Box 44).

As well as working hours and childcare arrangements, the culture of an organisation towards new mothers and fathers also has an effect. According to a study on new parents (Borrill and Kidd, 1994), women had more difficulty discussing their altered circumstances and expectations with their managers than men, due to feelings of insecurity, anxiety and mistrust of the organisation. To promote more even task sharing in the home, action is also needed to change traditional perceptions of roles of male and female parents. Artazcoz (2001) suggests that gender-sensitive education activities in schools, workplaces and mass media are needed in order to increase the consciousness of sharing domestic roles and responsibilities.

While positive action to reconcile work and family life is clearly particularly important for women, improved family-friendly work policies benefit both parents and research suggests that men are concerned about this too (O'Brien and Shemilt, 2003). In the 'traditional' work model, where men work full-time and are frequently

Box 45: Men want work–life balance too

A survey and literature review for the Equal Opportunities Commission in the UK found high support for work–life balance among fathers as well as mothers. Fathers do approximately one third of childcare and this contribution has risen significantly since the 1970s. However, fathers have lower expectations of family-friendly working practices being available to them personally and are also less likely to take advantage of those that are in place. The report suggests that, in particular, a culture of working long hours and men's higher earnings (meaning that fathers are less likely to cut their working hours than lower earning mothers) prevent them from taking a more equal share of caring. The report also cites the experience of the Nordic countries where governments have found that fathers are more likely to take parental leave under four key conditions: where it includes a quota designated for fathers, where there is high wage compensation, where there is flexibility in the way leave can be used by couples and where provision of paternity leave for men is publicised through government awareness campaigns.

Source: O'Brien and Shemilt (2003).

A study of the 250 largest companies in Sweden suggests that there are four different attitudes where a male employee wishes to participate more in childcare at home: companies with explicitly positive attitudes encourage men to take paternity leave and feel that they are partly responsible for giving men more chances to combine work and family; the second type of company appears positive to paternity leave, but in practice is only positive when the company sees clear benefits for itself; the third type of company has 'neutral' attitudes; and the fourth practises passive resistance. A connection was also found between company culture and male employees' take-up of parental leave, as well as the gender composition of the workforce. Male employees in human resources-oriented companies took parental leave more often than male employees in companies with more production-oriented values. Also, the positive attitudes of colleagues and superiors towards parental leave had an influence on the rate of men taking parental leave. Companies employing mostly women were more positive towards men taking paternity leave than companies employing mostly men.

Source: Hwang (1999b).

expected to work long hours, many complain that they hardly see their children growing up. Also, men's long working hours are a block to the aim of furthering their participation in their family roles (Kauppinen, 1999a; see also Kauppinen and Kandolin, 1998). Therefore, 'work–family' policy debates should address men's schedules as well as women's. Furthermore, improving men's work schedules would enable more women to enter and progress in male-dominated employment areas as well as facilitating more men becoming involved in childcare and other household tasks. Further investigation is also needed into what women and men consider to be a 'compatible' work schedule and the reasons why (Fagan and Burchell, 2002).

Finally, better sharing of domestic and caring work between women and men would improve the quality of women's working life and facilitate their participation in the job market, and it could also bring positive changes for men as well. Based on its research findings, the European Foundation (2002c) states 'public authorities as well as social partners have a key role to play in developing measures to promote the reconciliation of work and family life. They are at the core of the tool for change.'

Occupational voice loss — an emerging risk

Keys points

- There is evidence of voice problems related to some jobs, such as teaching and speaking on the telephone.
- Women may be more at risk than men due to job segregation.
- Good practices for prevention exist.
- More research, awareness-raising and preventive actions are needed.

There is a small but growing amount of information and evidence about links between job and voice-related problems. The medical term for voice problems is dysphonia. The condition is not just an inability to speak but also includes pain, tension, croakiness, irritating cough, inability to modulate, poor or no vocal power and breathing difficulties (HELA, 2001). People who use the voice extensively will be at increased risk of suffering continuous vibratory stresses of the laryngeal mucosa (causing swelling and injuries such as mucosal haemorrhage) and developing voice fold polyps and nodules (Dikkers and Nikkels, 1995).

Apart from singers, incidents have been documented among teachers (for example, see Cortazar Lopez et al., 2002; Harisinghani, 2000), daycare staff (for example, Sala et al., 1999) and jobs requiring high telephone use (such work includes telephonists, telephone sales and call centre work). Because of job segregation, work-related voice problems may affect women more than men. For example, in a study of 430 patients presenting themselves with voice complaints at the doctor's surgery, a 14 % incidence of mucosal disease was found in women and 10 % in men (Bastian and Thomas, 2000).

Suggested preventive measures include: design of workplaces to reduce noise; breaks; actions to reduce stress at work; attention to air humidity; access to drinks; avoidance of talking with a cold; training to speak with relaxed voice-box muscles, using the larynx's optimum pitch and in voice projection; training and awareness to recognise the symptoms early and what types of voice use to avoid (see Box 46).

Further investigation is recommended, including preventive measures in the classroom. Possible gender differences that could be investigated include: different male and female voice pitch — it could be that, in some circumstances, some pitches are easier to hear than others, so these people do not need to raise their voices as much — and job segregation (within teaching, a higher percentage of female teachers work

Box 46: Voice health at work for call handlers

Call handlers spend a larger proportion of their working day speaking on the telephone than many employees in typical office jobs. In a study of call handlers, participants made comments about the length of scripts sometimes straining their voices. Concern was also expressed about the frequency and duration of throat infections. However, the risk of them experiencing problems with their voices can be reduced if good practices are followed:

- To reduce the risk of straining the throat, opening greeting scripts should be broken into shorter segments, giving call handlers frequent micro-breaks while callers respond to their questions.
- Allow call handlers to drink at their workstations to ensure their throats are adequately lubricated.
- Call handlers should be encouraged to drink water or caffeine-free soft drinks to maintain hydration rather than tea, coffee or soft drinks containing caffeine, which are diuretics.
- Stretching the neck and shoulders relieves tension. These exercises can be done at the workstation as well as during breaks. A prompt that appears on the screen from time to time may be a helpful reminder for call handlers to do these stretches.
- The risk of voice problems is greater when suffering from a cold. Assigning staff in these circumstances to tasks that do not involve speaking on the telephone reduces this risk.
- Voice tubes can become blocked with food, make-up and dust, and this compromises the effectiveness of microphones. Train call handlers to clean the voice tubes in order to optimise the volume of the transmitted signals and avoid the risk of frustrated callers and strained voices.
- Provide call centre employees with information on the risk of dysphonia, the various symptoms of the condition and how this risk can be reduced.
- The introduction of any good practice for voice health that may substantially affect the health and safety of call handlers should be done in consultation with call handlers and their representatives.

Sources: HELA (2001). See also BIFU, 'Voice Care Network'.

---

with younger age groups compared to men. Are some age groups noisier than others?). Awareness-raising and preventive actions are also needed.

Appearances

Some jobs require employees to wear specific clothes and the requirements of dress should not conflict with health and safety. While there may be such requirements placed on both men and women workers, they may affect women more.

Uniforms should be designed with the job in mind. For example, in some countries, the design of nurses' uniforms frequently used to be quite tight fitting, with a straight skirt. Some were also expected to wear a cap. Both dresses and headgear can inhibit movement for carrying out work tasks safely, such as manual handling, while trousers or culottes are the most practical solution (Rogers and Salvage, 1988). At least, in healthcare jobs, there is general recognition of this, for example see Health and Safety Commission (1998), and suitability of clothing is mentioned in the European directive on manual handling at work. But, in other jobs, for example in the case of shop staff who may also have to carry out lifting tasks, there may be less appreciation of the additional risk. As has been referred to, clothing requirements can be an issue for women going through the menopause and who are subject to hot flushes, and they should be able to wear layered and loose clothing (Paul, 2003).

Dress requirements for work may also specify the type of shoes to be worn. Unsuitable footwear can contribute to slipping and tripping accidents or it may just be uncomfortable if the job involves a lot of standing or walking. In some jobs, such as in retail shoe shops, staff may find themselves wearing shoes with heels or slippery soles to climb step ladders to reach stock. Fashion models if required to wear extremely high, narrow heeled or platform shoes may be at risk of tripping and falling and of twisting or breaking an ankle. Workers need to have footwear that is suitable for their working environment, taking account of the type of work, floor surface, typical floor conditions and the slip-resistant properties of the sole (European Agency, 2001c).

In some jobs, there may be a certain pressure, even if not explicit, to meet physical appearance requirements which could be prejudicial to health if required on a long-term basis. Actors may gain or lose weight for a particular part in the short term, or, in general, both female and male actors can be under pressure to maintain a youthful appearance in order to keep working, although the pressure is probably higher on women. The requirements on models regarding aspects of physical appearance, such as to maintain a certain body weight, can be demanding, but other parts of the fashion industry, including high street garment stores, may put pressure on female staff concerning their weight, in the belief that model-like staff will sell more clothes. Previously, there were expectations that airhostesses should conform to certain appearance standards, such as weight, although this has now been relaxed.

The health and safety of women in agriculture

Key points

- The agricultural sector is still a significant employer of women in some Member States and in the candidate countries.
- Many more women contribute as wives, partners or family members. These workers may not be covered by occupational safety and health legislation.
- Women are concentrated in the elementary jobs in the agricultural sector.

- Although the accident rate is higher for men, both men and women report high levels of accidents in the sector.

The European Parliament (2002) has, in its discussion of a Community strategy on safety and health at work, highlighted the need to pay closer attention to women working in the agricultural sector.

The agricultural sector is a significant employer of both women and men in the EU. Agriculture is the fifth highest employer of men in the EU, some 5 % of the total workforce, and the seventh highest employer of women, accounting for 3 % of women in employment. In Greece and Portugal, agriculture is still the main employer of women (Dupré, 2002). In the candidate countries, the sector is much more important, accounting for 21 % of employed people compared to 5 % in the EU according to the European Foundation (Paoli, 2002), with the rate for Romania at 45 %. The sector is more important in the candidate countries than in the Member States, including for the employment of women. Many women will also contribute as wives or partners of farmers, and temporary, casual seasonal work, for example during the harvesting season, is probably an important feature of women's work in this sector. Traditionally, a farm in Europe is a family concern worked basically by a couple, with the wife helping her husband with numerous daily tasks (Fremont, 2001). In the United States, women's participation in agriculture is increasing, for example with more daughters entering family businesses (McCoy et al., 2002). Of family workers in agriculture in the EU, 38 % are women (Linares, 2003, and see Box 47).

There is likely to be considerable task segregation by gender; for example, traditionally on some farms men picked apples (involving working up a ladder), and women harvested soft fruit from the ground (involving work in a stooping position). The apple pickers earned more.

Women are concentrated in elementary jobs in the agricultural sector (European Foundation, 2002c). Women farm managers are generally older and have had a lower level of agricultural training than men (Fremont, 2001).

> **Box 47**
>
> Women family agricultural workers are generally:
> - 38 % of family workers;
> - older than men;
> - working part-time (86 %);
> - the spouse of the holder (three in four spouses are women; almost one holder in four is a woman);
> - managing smallholdings.
>
> *Source*: Linares (2003), 'Women and men in agriculture: a statistical look at the family labour force', *Statistics in Focus*, Theme 5: Agriculture and fisheries, 4/2003, Eurostat.

According to Eurostat figures, adjusted for hours worked, in the EU, men only have around a 1.5 times higher accident rate than women in agriculture. This compares to around 2.33 times higher in manufacturing and 3.33 times higher in construction (see Table 1 and Figure 3 in Annex 7). According to the third European survey on working conditions, agricultural workers report a high level of back injuries (57 %). They are high among the occupations likely to suffer accidents — 13 % report suffering accidents compared to a 7 % average. Up to 41 % in the agricultural and fishing sector think that their health and safety is at risk because of their work (Paoli and Merllié, 2001).

Hazards on the farm include exposure to pesticides (that can have reproductive health effects), exposure to diseases transmitted by animals, manual handling and repetitive work in awkward postures, exposure to harmful rays of the sun and cold and wet weather conditions, use of dangerous machinery, etc. (see also Annex 8).

According to McCoy et al. (2002), to date, few studies have examined work-related injuries among farm women. Even less is known about the extent to which occupational risks are recognised when women seek medical care. They suggest that differences in size and stature, increased physical strain, and low maximal oxygen uptake may predispose women to ergonomic-related injuries. Again, the design of tools and equipment for a diverse population will be an issue and examination of the real jobs carried out by women agricultural workers in relation to occupational safety and health is needed.

The many women who are partners in family-run farming businesses and not employees may fall outside the scope of safety and health legislation. Some Member States, such as the UK, have decided to treat such farm workers as employees for the application of safety and health law, and have covered the issue in sector guidance (Health and Safety Commission, 1999).

Health and safety in the fisheries sector

The fisheries sector is very male dominated, especially work on seagoing fishing vessels where women only make up 3 % of the workforce, but they are more involved in other parts such as: aquaculture; processing — where they are over-represented, predominating in the low-grade, unskilled jobs; management and administration; and, as in the agricultural sector, informal support to seagoing spouses and partners, where they play an important, yet undervalued role.

One EU study for the European Commission looked at equality issues among women in the sector, although it did not cover occupational safety and health issues. It found that women feel unwelcome in the seagoing sector, but have little interest in participating for reasons such as health, safety and welfare issues including the heavy physical work involved, need for extra facilities on board, dangerous nature of the work, problem of mixed crews on cramped boats and 'rough' male culture on board. In other parts of the sector where women are more present, significant discrimination was found. The study found that vocational training and education is likely to be key to improving women's position and prospects in the sector. Safety and health issues were not mentioned as topics for training, but it would seem logical to include them in any training initiatives. Similar to the situation in the agricultural sector, many women support seagoing spouses and partners, including management and administration functions, communication, marketing and practical back-up. The study also recommended training for these women, including safety at sea (MacAlister Elliott and Partners, 2002). The European Community strategy on safety and health at work 2002–06 also mentions the importance of strengthening the role of women in improving working and safety conditions within the fishing industry.

The health and safety of domestic workers and homeworkers

Key points

- Both domestic and homeworkers are a female-dominated workforce and a significant employment area for women.
- A diverse range of jobs and tasks may be carried out in both areas, exposing them to a variety of hazards.
- Poor access to occupational safety and health advice, information and training, isolation and an informal and precarious employment situation are among the factors that could increase exposure to hazards for this group of workers compared to workers carrying out similar tasks on business premises.
- Many homeworkers are technically self-employed so do not come under the scope of the European legislative framework, and domestic servants are specifically excluded.

The European Parliament (2002) has, in its discussion of a Community strategy on safety and health at work, highlighted the need to look at the occupational safety and health protection of domestic workers and homeworkers. This female-dominated group falls outside the scope of the European occupational safety and health legislative framework. The work is characterised by a lack of employment rights and being casual and low-paid work. Much of the work is undeclared. According to Eurostat figures, 2 % of employed women work in private households, which represents the eighth highest employment sector for women (Eurostat, 2002b). Some women may choose this type of work as its flexibility suits their family obligations, for example.

Domestic and household workers

Paid domestic work is diverse in nature and will include cleaning, laundry work, cooking, caring for children and the elderly and infirm, helping at family celebrations, gardening and home maintenance (Cancedda, 2001; Lutz, undated) and exposure to hazards associated with this work. Included among these hazards are toxic chemicals, heavy lifting and repetitive movements, awkward postures and working while standing (Valls-Llobet et al., 1999). These women workers may also be more at risk of intimidation and harassment by the people they work for, due to their vulnerable and isolated situation. They may work for two hours a day in a home, but often work in various different homes, or they may be on-call 24 hours and live on the premises of those they work for. Household work is one of the largest employment areas for immigrant women (Cancedda, 2001) and they may be even more vulnerable — there have even been cases reported where women have been found working virtually in a state of slavery, as house prisoners.

Although there is not much information available on this sector, some information that does exist suggests that it is high risk, for example Vogel (2002) quotes Belgian work accident data from 1998 that 'shows an overall severity rate well above the private sector average (12.10 per 1 000 against 2.18)'.

In a report for the European Commission concerning immigrant domestic workers, Anderson and Phizacklea (1997) list the following as common problems faced by domestic workers:
- unpaid hours;
- low income, often less than the minimum wage; denial of wages in cases of dismissal following the trial or probation periods;
- refusal by employers to arrange legal resident status (for tax reasons etc.);
- control and sexual harassment;
- pressure to do additional work (for friends and colleagues);
- excessive workloads, especially where in addition to caring for children and elderly people they are responsible for all other household chores;
- very intimate relationships between the domestic helpers and their employers.

Domestic workers are even less likely than women doing similar tasks within enterprises to receive information, instruction and training on safety and health hazards. Nor are their employers likely to receive information about risks to domestic workers and their prevention. A quick search of the Internet found information for employers about vetting the criminal record of domestic workers and even the possibility of diseases being passed on by immigrant domestic workers, but nothing about ensuring the safety of those they may employ. The fact that this group of workers falls outside the occupational safety and health system is probably one of the reasons why there is also little occupational safety and health information available about or for this group. In addition, they are a more or less unrepresented group of workers, for example by trade unions.

Australia is one country that has made changes to its employment and compensation legislation to include domestic servants. In 1998, domestic workers became covered under the Australian workers' compensation scheme and in 2002 sections of the Labour Act were amended to include domestic servants — Labour Act (Chapter 93), 'Labour (domestic servants) rules'.

Box 48

A survey of household work was conducted by the European Foundation (Cancedda, 2001) in eight Member States. The results suggest that work in household services, particularly in childcare and care of the elderly, can be rewarding and satisfying for the often highly motivated workers. The survey found that the work of household workers exposes them to mental and physical strain, and they often have inadequate support. Workers in household services are mainly women and the majority of jobs created in this area are low status, low skilled and poorly paid. Part-time work is common in household services, and because the hourly wages are low, workers often see part-time work as a limitation rather than a choice. Regarding social protection, there are large differences between countries. Some countries have tried measures that fall halfway between self-employment and employment, for example through service agencies. This can provide a steady income and ensure a minimum of social benefits (Cancedda, 2001).

Box 49: Domestic workers in Spain

Spain's domestic service sector is female-dominated and is increasingly becoming the main form of integrating female immigrants into the labour market. Domestic workers are covered by special laws concerning employment and social security, which are inferior in many respects to those enjoyed by other workers. There is research evidence that as many as 60 % of the domestic workers are in the unregistered, underground economy and that 35 % of those registered are immigrants. Employment is often irregular by nature, and both payment and working conditions are poor. Particularly those who live on the premises earn little, as accommodation, board, etc., are considered as payment in kind. There are almost 400 000 homes with domestic employees in Spain. They are a particularly unrepresented group. However, in 2002, Spanish trade unions raised the issue of their problems (Albarracan, 2002).

Homeworkers

Working from home is not a new phenomenon. However, the use of information technology has increased the possibilities, and this sub-group are often known as teleworkers. Homeworkers carry out a range of activities including sewing and knitting, packing, assembly, soldering and telesales. Technological changes have meant that skilled workers are now more likely to work from home. Homeworkers are therefore far from being a homogeneous group and the health and safety issues they face vary. However, a large proportion of them are women. One study on what is known about homeworkers and their occupational safety and health has been carried out in Great Britain for the national OSH authority, in response to concerns that a proportion of homeworkers could be working under conditions that were unacceptable 'in usual circumstances' (O'Hara, 2002). It was based on a literature review, focus group discussions with homeworkers and interviews with employers. It was carried out to shed light on the numbers, trends, reasons for homeworking and risk factors facing homeworkers in Great Britain and to provide recommendations for further research into the health and safety of homeworkers.

The main findings of the report included the following:

- In Great Britain, homeworkers account for 2.3 % of the employed workforce. Census and labour force survey data reveal a considerable increase in the number of people working at home since the early 1980s.
- Women constitute the majority of those working mainly at home, though men are more likely to work at home on an occasional basis. Women homeworkers are more likely to work in manufacturing than non-homeworkers. The opposite appears to be the case for men.
- Ethnic minorities were found to be under-represented in homeworking compared to the employed workforce as a whole, but they were over-represented in manual and low-paid homework.
- The main industry sectors for homework activities were business services and manufacturing and the main occupations identified were sewing, assembly and packing and non-manual occupations such as clerical, secretarial and administrative work. The majority of homework suppliers are small firms.
- The main reasons for working at home for employees were childcare, financial and flexibility, and the main disadvantages were poor pay, isolation, mess and irregularity of work. The main reasons identified for employing homeworkers included flexibility/dealing with fluctuating workflows, reduced costs, restricted space, and solving childcare problems, and the main disadvantages were difficulty with supervision and reduced contact with staff.
- Environmental hazards caused by homework included: lack of space, dirt, smell, noise, electrical, and fire. Hazards perceived as causing accidents and ill health included poor seating, repetitive work, manual handling and working with substances such as solder, glues and paints. The main health problems experienced by homeworkers were musculoskeletal pain, eye strain, headaches and mental strain.
- The research studies provide evidence of accidents affecting homeworkers and others in their home, including children. Accidents and health problems often go unreported to the company supplying the work. Even those homeworkers who sought medical treatment had not always informed the health professionals that their problem was work-related.
- Levels of awareness of health and safety issues appeared to be quite poor as was access to health and safety information, equipment and training. Homeworkers had not seen

official guidance and were not familiar with health and safety legislation relevant to homeworking. Risk assessments were not being carried out. There is a high level of under-reporting of accidents.
- Many homeworkers were in favour of the labour inspectorate focusing more attention on homeworking and the possibility of visits to homes. In relation to the kind of guidance information that would be useful, some expressed a preference for brief leaflets, which are more specific to the different types of activities. Translation of guidance and videos or other pictorial formats were also mentioned.
- Both homeworkers and their employers appeared to be confused over employment status and its implications for health and safety provision.

The report's recommendations included the following:

- Further research, for example to: obtain more recent data; obtain baseline information; look at differences between homeworkers in different sectors and compare those who work at home some of the time with those who work at home all of the time and those who do homework with employer-based workers doing the same jobs; investigate which kinds of tools and interventions would be most appropriate; and explore issues such as homeworkers working from home because health problems or disability make it difficult for them to go out to work.
- Carrying out sector-based case studies of good practice in health and safety for homeworkers, to be used to develop guidance on identifying risks and appropriate control measures. The development of guidance should include piloting and address the need for translation into languages other than English.
- Investigating a reliable system of recording accident and ill-health data among this group.
- Identifying more effective means of communicating on occupational safety and health to homeworkers and their employers.
- Exploring the issue of confusion over the coverage of homeworkers with self-employed status by occupational safety and health regulations.

Box 50: EU social dialogue agreement on teleworking includes safety and health

The subgroup of more recent homeworkers who use information technology, whereby often their home becomes their office, are known as teleworkers. An EU-level framework agreement on teleworking between the social partners (employer and trade union organisations UNICE/UEAPME, CEEP and ETUC) on teleworking includes health and safety elements. This is a good example of mainstreaming occupational safety and health into other employment areas. There is a three-year implementation period for this voluntary agreement, with effect from July 2002. For further information, see:

- European Commission, Employment and Social Affairs web site (b), 'Teleworking agreement';
- European Commission, Employment and Social Affairs (2002b), *Social Agenda* 3. 2002;
- InfoBASE web site.

Occupational safety and health of women working in 'non-traditional' areas

The National Institute for Occupational Safety and Health (NIOSH), in the United States, is carrying out research on the health and safety of women in non-traditional employment areas. It points out that 'women in non-traditional employment may face health and safety risks due to the equipment and clothing provided to them at their workplace. Personal protective equipment and clothing are often designed for average-sized men. The protective function of equipment and clothing (such as respirators, work gloves and work boots) may be reduced when they do not fit female workers properly' (NIOSH factsheet). They may also hinder carrying out the work safely and properly.

NIOSH estimates that women's participation in the US construction sector will increase over the next 10–15 years. As a response to this issue, it carried out interviews and surveys with women in the sector and, based on this, produced a list of issues and ideas for providing safety and health protection for a diverse construction workforce. The issues highlighted were training, personal protective equipment, sanitary facilities, workplace culture, ergonomics and reproductive hazards (NIOSH, 1999a).

Women who work in non-traditional employment settings may also face specific types of stressors. For instance, they may be exposed to sexual harassment and gender-based discrimination.

It is also important that working conditions are not a barrier to women entering 'non-traditional' areas of employment.

> **Box 51: Women in transport — an example of 'non-traditional' employment**
>
> Although previously uncommon, jobs that involve travel are not a new area of employment for women. In the mid-19th century, for example, Scottish female fish workers used to travel hundreds of miles every year to follow and gut the herring (Kumpulainen, 2000). However, in recent decades, the proportion of women working in occupations that call for physical mobility has been growing rapidly. One of the biggest problems in this kind of work is the reconciliation of work and family.
>
> Train and lorry drivers, pilots, and other transport workers have traditionally been men, but the number of women workers is gradually growing. The International Transport Workers' Federation carried out a survey of their members in civil aviation, docks, fisheries, inland navigation, railways, road transport, seafaring and tourism services. The results show that women workers in the transport sector are subjected to various forms of discrimination: the standards and attitudes concerning recruitment, pay and promotion, maternity, marital status and company image are more favourable towards men. The transport industry is also known to have a high risk of violence (International Transport Workers' Federation, 2000, 2002).

Working in SMEs

Key points

- While men are more likely to be self-employed or run small businesses than women, many women will be involved as wives or partners, and women are over-represented as employees.
- There are examples of schemes aimed at risk prevention in SMEs that cover female-dominated work areas, but it will be important to ensure that resources are going towards both risks to women and men in SMEs.
- Initiatives aimed at providing worker representative coverage on safety matters in SMEs would be beneficial to the many women working in SMEs who have no form of representation.

According to the European survey on working conditions, of the 17 % of the working population who are self-employed, only one in three is a woman. However, women are over-represented as employees in small, private firms, while men are more likely to work in medium-sized and large private companies (Fagan and Burchell, 2002). Small firms have a poorer health and safety record compared to medium-sized and large firms, their employees are less likely to receive training and they are less likely to be members of a trade union.

As in the agricultural sector, many more women will support their husbands or partners who run SMEs, or are self-employed, without being visible or recognised as workers, so they are also ignored for health and safety purposes. One French initiative has recognised that even among small businesses in the construction sector, spouses and partners can play a significant role: a professional association for prevention in the building industry, together with the construction federation and the construction industry's women's group, developed training resources for spouses and partners involved in very small enterprises in the construction industry, to enable them to identify risks and prevention measures. As many of the women carry out personal tasks, they can also play a very important role in the enterprises' management, including in the enterprises' prevention policy (European Agency, 2003a).

Member States have been introducing schemes to raise awareness and provide practical support and resources to SMEs to help them manage health and safety risk prevention (for example, see European Agency, 2003c). Partnerships of OSH organisations, social partners and non-governmental organisations (NGOs) have been carrying out a variety of SME-focused OSH projects (for example, see European Agency, 2003a,

---

Box 52
Examples of European transnational and national SME projects to improve health and safety

A collaborative approach in their development and active dissemination of the resulting information, for example through targeted distribution of free resources, and using sector organisations, is an important element of these projects:

- Cross-border collaboration in the hotel and catering industry — creation of networks and partnerships.
- Avoiding bloodborne diseases — CD-ROM, web site and leaflet.
- Health and safety training for spouses and partners in the construction sector.
- Voluntary sector — health and safety in charity shops: information pack and workbook.

Sources: European Agency (2003a, 2003b).

Examples of EU sectoral-level programmes for SMEs targeting female-dominated workplaces

- For the dry-cleaning industry: a self-monitoring system, technical support and facilitation of information exchange, underpinned by regular inspections and an award system (trade federation in cooperation with an OSH organisation — Germany).

Source: European Agency (2003c).

- Preventing musculoskeletal disorders in crèches (in development).
- Risk evaluation and prevention in cleaning companies (in development).
- Textile and garment industries' campaign (in development).
- Stress prevention for psychiatric therapy and rehabilitation workers (in development).
- Stress management in hotel, restaurant and café trades (in development).
- Safety management in the tourism industry (in development).
- Partnership agreement between private companies and temporary employment agencies (in development).
- Developing best practices in bakery, patisserie and confectionary industries (in development).

- For the retail sector: a tool for identifying risks and implementing practical solutions (national OSH institute, Netherlands).
- For the ceramics industry: a national action plan for risk prevention through integration of occupational risk prevention into daily management, covering research, information and training (national OSH institute in cooperation with social partners, Portugal).

2003b). While there are examples of actions that cover female-dominated work areas, it will be important to ensure that project resources are going towards both risk prevention to women and men workers.

Schemes or systems for worker safety representatives or advisers to operate on a regional basis, with the rights to consultation and to provide representation in a number of smaller workplaces, could be particularly useful to unrepresented women employees in smaller workplaces. Some Member States, which do not have provisions for union safety representatives to operate on a regional basis, have been successfully piloting such an approach, including in the hospitality and voluntary sectors, which are high-employment sectors for women (for example, see York Consulting, 2003).

Occupational safety and health of commercial sex workers

For many women in the EU, sex work is an occupation. However, sex workers are often overlooked in activities on occupational health and safety. This can be because their work is often illegal and not covered by occupational safety and health or other employment legislation. Research and interventions for this group also have most often focused on sexual health issues, and, more recently, work-related violence, to the neglect of other issues such as broader physical or mental health issues. Hazards faced by sex workers include: sexual health hazards, including sexually transmissible diseases and resultant mortality, infertility and gynaecological problems, violence and drug and alcohol abuse problems and mental health hazards, with stigma and powerlessness as contributing factors (Plumridge, 1999).

Plumridge also points out that the occupational health of sex workers is influenced by local legislation, customs and contexts, as the example in Box 53 from the Netherlands also illustrates, and that the most crucial determinant of health is the degree of control they have over their working lives. Those working outside the law are at much greater risk.

Some OSH authorities have produced guidance in this area. For example, WorkCover, the Health and Safety Authority in New South Wales, Australia, has produced health and safety guidelines for brothels (WorkCover, 2001). The issues affecting sex workers were also addressed in papers presented at the third International Congress on Women, Work and Health, 2002. Conclusions from the congress were that: those involved in programmes and policy related to the health and working conditions of sex workers should cover the whole range of relevant occupational safety and health issues; health interventions should be responsive to what sex workers really think and want from interventions; and it is important that interventions are carried out in a context free from moral or cultural judgment. Finally, it is worth commenting that there is probably even less information available about male prostitutes.

Box 53: Prostitution and legislation

Prostitution and brothels became legal businesses in the Netherlands in 2000. By legalising prostitution, the Dutch Government aims to protect and improve the position of sex workers and prevent illegal immigrants from going into prostitution. Sex club owners now have official permits and licences to run their businesses as long as they follow the legislation, including detailed guidelines and standards concerning health and safety. The legislation makes a distinction between voluntary and involuntary prostitution, which is aimed at enabling the authorities to regulate the sex industry, abolish the modern 'sex slave' trade and sexual abuse of minors, and combat the criminal side effects of prostitution, such as money laundering and drug dealing. However, only women from within the EU are allowed work permits, as the law governing entry of non-EU labourers into the country does not recognise prostitution (Schippers, 2002).

The work-related health of older women

Key points

- Older workers are not a homogeneous group, and gender differences do exist.
- There is not much specific literature on the health at work of older women as a group.
- Policy on the ageing workforce should take account of gender.

Demographic changes mean that the average age of the workforce is increasing. This means that older women, over the age of 45, now hold an important place in the labour market. For both employers and the economy as a whole, as well as for the well-being of older women themselves, it is important that work is not impacting negatively on their health. It is now recognised that the health and safety of older workers as a group merits attention, and ageing workers are featured among the priorities in the European Community strategy on health and safety at work 2002–06. This strategy includes the objective that 'preventive measures should take account of the age factor and particularly target young workers and ageing workers'. In addition, it has also been shown that early retirement is more common in enterprises with poor quality working environments (Forss et al., 2001; Hakola, 2002). In general, action is needed to reorganise the working environment to prevent premature wear and tear and allow people to remain longer in work (European Foundation, 2002a).

Although much more attention is now being paid in the scientific literature and in terms of interventions related to issues for older workers in general (for example, European Foundation, 2002a, 2002b), there is very little specific literature on the health at work of older women as a group. One source of information is a literature review by Doyal (2002). Among the issues for the health of older working women in the 45–64 age group highlighted in the study are:

older workers are not a homogeneous group and there will be gender-related differences within the group; menopause-related issues; job segregation with older women being even more likely to be concentrated in the traditional 'female' occupations with lower rewards and status, as they are less likely to have benefited from the equal opportunities strategies of recent years; as they grow older many women move from part-time to full-time work; longer work exposure to hazards due to the length of time working; musculoskeletal problems and stress being the most commonly reported symptoms by this group; and their domestic workload (there may be a more uneven sharing of domestic responsibilities among older couples than younger ones), including their often greater role as carers of elderly relatives.

The study by Doyal (op. cit.) makes a number of recommendations for stakeholders — employers, trade unions, researchers and governments — concerned with the health of older women workers. These are focused on taking an approach that is both age-sensitive and gender-sensitive, to tackle both age and gender discrimination. Work-

> **Box 54: Support for carers of elderly relatives**
>
> Older women workers are often carers of elderly relatives. Employed carers in Germany report high levels of stress and very often experience work interruptions or have to miss work altogether; they are prone to financial loss, for example, due to unpaid time off; and they often miss business meetings and training opportunities and lose out on promotions. The strongest negative consequence for employers is the loss of qualified, committed and experienced employees.
>
> Siemens is one German company that has implemented a policy for carers who are in employment. The measures involve flexibility in working time and place of work and leave options, as well as an information and referral service, and counselling and advice in connection with company-based social work measures.
>
> *Source:* Naegele (1999).

place OSH recommendations include that risk-assessment and management measures take account of both age and gender issues, that older women workers are consulted, and that appropriate occupational health services and support for carers are provided. Research recommendations include paying attention to age and gender in the design of studies, and involving women workers in the design and implementation of studies. OSH recommendations for governments include: gathering more age and gender-related data; promoting support services for employed carers; developing an age-sensitive and gender-sensitive OSH approach; and for the European Commission, in particular, to take account of the needs of older women workers when planning its OSH policies and programmes. The age-sensitive and gender-sensitive OSH recommendations for trade unions include: awareness raising; negotiating; training of worker representatives and involvement of older women workers.

Migrant women workers

Many workers, including women, migrate to find work or better work, or because they have become refugees. Workers may migrate between countries within the EU, or come from outside the EU. Kane (1999b) reports that international migrant women are to be found 'disproportionately in low-paid and poorly unionised factories where they may experience inadequate lighting, ventilation and safety precautions. They are also concentrated in the non-formal sector where, for example, women street vendors or leaflet distributors may face exposure to sunburn and melanomas: working conditions amongst such women are seldom studied. Women migrants are also disproportionately employed in homework and paid — but poorly — by the piece; they often work very long hours and the pressure increases the risk of accidental injuries from, for example, sewing machines.' They may also be drawn into prostitution. Language barriers may prevent migrants from understanding occupational health and safety information and training, raising issues or obtaining help and also create difficulties for them in obtaining help from healthcare systems. The invisibility of women in migration data makes it difficult to ascertain much about them (United Nations, 1995).

Occupational hazards and gender worldwide

This report concentrates on looking at working conditions and gender within the EU and the candidate countries. However, with the increase in globalisation, it becomes increasingly appropriate to consider the situation elsewhere in the world, including developing countries. As well as comparing the situation regarding gender and working conditions to that in Europe, it is useful to consider what strategies have been adopted for prevention. At the international level, the ILO and others have carried out activities to analyse gender issues in relation to occupational hazards. Many of the findings correspond to the situation at the European level described in this report, and some further information is given in Annex 8.

Different jobs, different exposures — implications of gender segregation

As has been seen in this section, different occupations, different tasks and different numbers of hours worked mean different exposure to hazards and different health outcomes. Many differences in working conditions are more closely related to sector and position in the occupational hierarchy than to gender per se (Fagan and Burchell, 2002). Fagan and Burchell found in their analysis of the European survey on working conditions data that overall gender differences were still present when comparisons were made within occupational status groups and, in some instances, the difference was more pronounced when it was averaged across all white-collar and blue-collar employment. For some working conditions, they found an interaction between gender and occupational status, where the gender pattern within a particular group contradicted the overall comparison for all employment. The examples they provide are that the average gender difference in ergonomic conditions was slight, but, among professionals, women were more exposed to ergonomic hazards and also the minority of women working in craft jobs were more exposed to ergonomic hazards than men in this occupational area. Similarly, they found that repetitive tasks were particularly prevalent in blue-collar jobs, and that among blue-collar workers the rate was highest among women. Regarding material and physical hazards, they found that in both blue-collar and white-collar jobs, men were more exposed to these risks than women, with the gender difference most pronounced in manual jobs. However, overall, exposure was highest for men in manual jobs, followed by women in manual jobs.

The presence of vertical and task segregation means that it cannot be assumed that women and men working in the same sector or even in the same job are carrying out the same tasks — i.e. exposed to the same hazards. This has implications for research and the analyses of accident and ill-health statistics by gender, as well as for risk assessments and interventions.

Messing and Stellman (1999) propose that the gender division of labour affects women's health in at least six ways:

- Women's jobs have specific characteristics (repetition, monotony, static effort, multiple simultaneous responsibilities) which may lead over time to changes in physical and mental health.
- Spaces, equipment and schedules designed in relation to the average male body and lifestyle may cause problems for women.
- Occupational segregation may result in health risks for women and men by causing task fragmentation, thereby increasing repetition and monotony.
- Sex-based job assignment may be vaunted as protecting the health of both sexes and thus distract from more effective occupational health-promotion practices.
- Discrimination is stressful in and of itself and may affect mental health.
- Part-time workers are excluded from many health-promoting benefits such as adequate sick leave and maternity leave.

Fagan and Burchell have summarised the relationship between gender, occupational status and working conditions found in the data from the third European survey on working conditions (see Annex 9). Also, Stellman and Lucas (2000) have summarised some of the main hazards that women can expect to encounter in the various domains of work in which they predominate (see Box 55). These include physical, chemical and biological hazards as well as psychosocial hazards.

Box 55: Examples of hazards in areas of women's work (after Stellman and Lucas, 2000)

Area of work	Biological hazards	Physical hazards	Chemical hazards	Stress
Motherhood and childcare	Infectious diseases (particularly respiratory)	Injuries associated with lifting and carrying	Household cleaning agents	Stress associated with caring occupations; burnout
Water, sanitation and cleanliness	Infectious diseases (particularly waterborne)	Injuries associated with lifting and carrying	Household cleaning agents	
Healthcare	Infectious diseases (especially airborne and bloodborne)	Injuries associated with lifting and carrying; ionising radiation	Cleaning, sterilising, and disinfecting agents; laboratory agents and drugs	Stress associated with caring occupations; burnout

Area of work	Biological hazards	Physical hazards	Chemical hazards	Stress
Food production	Infectious diseases (especially animal-borne and those associated with moulds, spores and other organic dusts)	Repetitive movements (e.g. in slaughter houses and meat-packing); knife wounds; cold temperatures; noise; microwaves	Pesticide residues; sterilising agents; sensitising spices and additives	Stress associated with repetitive assembly line work
Food processing, catering and service	Infectious diseases (from contact with public); dermatitis	Injuries associated with lifting and carrying; wet hands; slipping and tripping; microwaves and heat	Passive smoking	Stress associated with dealing with the public; sexual harassment
Textiles and clothing	Organic dusts	Noise; repetitive movements	Formaldehyde in permanent presses; dyes; dust	Stress associated with assembly line work
Fuel and shelter		Injuries associated with lifting and carrying; exposure to elements; cuts and bruises from collecting and transporting materials	Polycyclic aromatic hydrocarbons from incomplete combustion of fuels	Stress associated with arduous labour
Commerce and distribution		Repetitive movements and eye strain, etc., associated with VDU use; etc.	Poor indoor air quality	Stress associated with dealing with the public
Education	Infectious diseases (particularly respiratory; measles)	Violence; prolonged standing; voice problems	Poor indoor air quality	Stress associated with caring occupations; burnout
Personal services other than child- or health-related care — anything from hairdressing to sex work	Infectious diseases (e.g. skin infections; AIDS and other sexually transmitted diseases)	Standing; lifting and carrying; violence	Chemical cleaning agents; hairdressing chemicals	Stress associated with caring occupations; burnout
Communications		Violence (journalism); repetitive movements (data entry); excessive sitting or standing	Poor indoor air quality	Electronic performance monitoring; fear of redundancy/unemployment
Light manufacturing		Repetitive movements (e.g. assembly line work); standing	Chemicals in microelectronics	Stress associated with repetitive assembly line work

Box 56: Photographic account of women's working conditions

Women represent over 33 % of the Italian workforce. INAIL, the Italian Workers' Compensation Authority, together with ANMIL, ran a photographic competition, aimed at illustrating women's lives. They then made a publication *Women at work* using pictures drawn from the competition. The aim is that these images should compliment scientific publications, publications on legal rights and guidelines, and conferences, exhibitions and training courses. The original competition was launched to coincide with International Women's Day. The photos cover women in traditional roles, but also pilots, sculptors, petrol pump workers, welders, restorers, fisherwomen and bricklayers.

Source: INAIL (2002).

4. ABSENTEEISM, DISABILITY, COMPENSATION AND REHABILITATION

In most European countries, employees are entitled to payment in the case of sickness absence, or to benefits in the case of occupational disability. However, in many countries, the condition has to be established as work-related in order to obtain the compensation. In the next sections, we examine sickness absence, occupational-related compensation arrangements and rehabilitation. It should be noted that information from different Member States, even within the EU, is often not directly comparable, due to the different national policies, infrastructures and practices, so only quite general comparisons can be made.

Sickness absence

Key points

- Women have more work-related sickness absence than men and slightly more non-work-related sickness absence than men.
- In the European labour force survey, younger women reported taking more non-work-related sick leave than younger men. This could be due to taking leave for children's sickness, especially as the trend is reduced in older workers. Other studies have not found a link between having children and women's sick leave rate.
- Some absence that is really related to work, such as minor infections among those working with small children, is usually treated as non-work-related.
- National legal and social policy context appears to have a large influence on sick leave and disability figures. Some Member States allow parents to take absence for children's sickness, for example.

The only cross-European data available for comparison of sick leave at present are self-report data. Self-reported sickness absence may reflect an underestimation of the real absence, since one often does not recall all days taken as sick leave over a period of, for example, one year. Making a judgment about whether sick leave is work- or non-work-related will also be affected by the level of awareness of the causes of particular health problems etc. However, in general, self-reported sickness absence is found to be correlated to the recorded sickness absence information, and is considered a valid indicator of relative differences and changes in the recorded sickness absence (e.g. Harrison and Schaffer, 1994; Rees, 1993).

Work-related sickness absence

According to the third European survey on working conditions, 10 % of women and men are taking sick leave due to health problems that they attribute to work. If a comparison is made by full-time and part-time workers, in each group women report more ill-health absence than men (part-timers are less likely to be absent than full-timers). The pattern of absence was found to be similar for both women and men (Fagan and Burchell, 2002. See also Table 7 in Annex 6). Causes of work-related absence have been dealt with in previous sections.

Non-work-related absence

Minor infections, such as common colds, are a significant cause of absence from work (for example, Vahtera et al., 1997). However, there is research evidence that a poor-quality working environment is as great a cause of absenteeism as, for example, a sedentary lifestyle characterised by obesity, lack of exercise, heavy smoking, or excessive consumption of alcohol (for example, Kivimäki et al., 1997).

According to the third European survey on working conditions, women report slightly more non-work-related sickness absence than men (Fagan and Burchell, 2002. See also Table 7 in Annex 6), although the difference appears greater if a comparison is made by full-time and part-time workers. Fagan and Burchell suggest that some of this difference could be related to gynaecological conditions but 'it is also because some workplace-related health problems may be particularly relevant in certain female-dominated occupations, such as (minor infections among) nurses and childcare workers, but are not recognised as "real" problems in traditional health and safety legislation'.

Some have suggested that family responsibilities may also explain some of women's increased absence, where, for example, the real reason for the absence is to look after a sick child (for example, see Grönkvist and Lagerlöf, 1999). However, there are studies that have not found higher rates of absence among women with children, except in the case of single mothers (for example, Mastekaasa, 2000; Gjesdal and Bratberg, 2002). On the other hand, women may compensate for having to take sick leave to look after children by not taking sick leave for their own illness. Results will also be influenced by national context. Very large variations in sick leave rates between countries are found, irrespective of gender. Differing labour and social security policies will have a strong effect. This includes how absence for employees and their children's sickness is treated by the employer, supervisors and collective agreements as well as by legislation and social insurance systems. Only some Member States provide provision for parents to take leave to care for sick children. See Annex 10 for a description of differences in Member State arrangements on absence and disability.

In any case, family responsibilities would only explain a part of the gender differences (Akyeampong, 1992), and other influences could be: women not taking sick leave if they are in temporary or insecure workplaces (Virtanen et al., 2001) and age — as older workers take longer to recover and younger women have entered the workforce more recently. Also, older women may be more likely to work part-time than younger women.

There appears to be both a work sector and type of job influence. For example, UK statistics and surveys have shown that public sector workplaces have higher absence rates than the private sector, large workplaces have a higher rate than small ones and manual workers have a higher rate of absence than non-manual workers (for example, Marmot et al., 1995).

Work-related stressors such as high job demands and uncertainty, fatigue and shift working may increase the risk of infections such as common colds, flu and gastric flu (Mohren, 2003). Various studies have found links between factors such as lack of control over work and having a lower position in the occupational hierarchy and higher sickness absence rates in both men and women (for example, Niedhammer et al., 1998; North et al., 1996).

There appears to be an interaction between age and gender regarding sickness absence. Analysis of data from the European labour force survey is hampered by the fact that questions on absenteeism are not the same in the different country surveys. This study did, however, result in the conclusion that, among workers of 20–54 years, women took more sickness absence than men in almost all the Member States covered by the survey. Amongst workers of 55–64 years, the trend was opposite: men being more absent than women in a majority of the countries. Again, the age differences could be partly due to younger women taking sick leave to care for children. And again, large differences between Member States were found and the factors of sex, age and skill level did not explain much of the differences in the total rates of sickness absence between the countries studied.

Compensation for occupational injury and ill health

Key points

- Compensation cover is better for risks found most often in male-dominated jobs.
- There is some evidence that women are less likely to be compensated than men for the same problems.

Gender differences in access to compensation for work-related injury and ill health are largely outside the scope of this report. There can be considerable variation between Member States (Gründemann and Van Vuuren, 1997) as official social security regulations for sickness absence and disability benefits vary between Member States (see Annex 10). Regarding compensation for work-related injuries, one common feature that has a gender impact is evident: it can be said in general that work-related injuries and damage from traditional and accident risks found most often in male-dominated jobs are better covered by compensation. This can be attributed to at least three reasons: first, the historical focus on these risk areas; second, where men were the most important or only wage earner, the aim was to ensure some compensation for the injured worker and his family; and, third, these types of injuries are more specific, with a single cause, and have a more evident work-related explanation. Therefore, there has been more willingness to compensate them (Grönkvist and Lagerlöf, 1999).

If a multi-factorial work exposure is present, as in many jobs dominated by women, the resulting disease is much less likely to be covered by industrial compensation arrangements, or, even if covered, it is much less likely to actually receive compensation. In some countries, mental health disorders are excluded, or only included with large restrictions as a legitimate reason for entering the disability benefit system. Similarly, in some countries, there are restrictions on the type of musculoskeletal disorders recognised under disability schemes. This approach lags behind preventive legislation in the EU that covers risks arising from multiple causes. In addition, some studies suggest that women are less likely to be compensated than men for the same problems (Lippel, 1993, 1996).

Rehabilitation into work

Key points

- There is a lack of research and statistical information regarding gender and rehabilitation into work following ill health.

- There appear to be some important barriers to women's participation in rehabilitation programmes.

Only a few studies have examined gender differences in rehabilitation into work. Some studies have found that, compared to men, fewer women are rehabilitated into work after a long spell of ill health (Veerman et al., 2001). Other findings indicate that rehabilitation rates are similar for men and women within the first year of absence (Houtman et al., 2002; Giezen, 1998). On the other hand, women have a higher risk of being diagnosed as disabled for work after the first year of sickness absence, whereas men are more often making progress towards return to work, and being provided with therapeutic support (Houtman et al., 2002).

A study of occupational health physicians suggests that they felt that rehabilitation into work was more important for men than for women (Vinke et al., 1999). Employers, too, appear to be more positive and more active towards the (at least partial) rehabilitation of men (Cuelenaere, 1997; Vinke et al., 1999). It appears that in particular less attention is paid to older women workers in vocational rehabilitation (Doyal, 2002).

Although the motivation to work does not appear to differ between men and women, women generally expect less from work at the outset of their career, and are more prepared to compromise their career to raise children (Jorna and Offers, 1991; Naber, 1991; van Schie, 1997) and this may affect their motivation to return to work. Other issues such as lower pay and lack of support from home may also be an influence. Another study found that women who cannot return to their original jobs due to injury or ill health are reoriented to a smaller list of new jobs than men (Lippel and Demers, 1996). The European Platform for Vocational Rehabilitation (2002) carried out a study to examine the participation and experiences of women in vocational rehabilitation in the various centres in the EU within its membership. It found various barriers to women's participation and made a number of conclusions and recommendations, including:

- there is a lack of national statistical information, especially information by gender;
- there is a need for more part-time courses and more flexible hours, as women, especially those with family responsibilities, find participation in full-time courses difficult;
- there is a need for more family-friendly facilities — childcare and/or financial support;
- more training is needed in 'typical women's work';
- the financial situation of women in vocational rehabilitation is often disadvantaged compared to men and can be a disincentive to participation;
- women may lack self-esteem to return to training and there is a need to improve career guidance advice for them; sexual harassment can be a problem, especially in centres with boarding facilities.

5.

GENDER ISSUES IN RESEARCH, LEGISLATION, RESOURCES, SERVICES AND PRACTICE

Introduction

Elements in occupational safety and health systems include: legislation, access to occupational safety and health resources and services, social policies and legislation; the system and format of industrial relations; what topics are prioritised for research or prevention; how research is carried out or statistics are collected; what information is available and how prevention activities are carried out. They all have an influence on what we know about occupational safety and health risks and how well they are prevented and therefore they should also be looked at from a gender perspective. The influence of factors outside the workplace on prevention within the workplace is shown in the model presented at the beginning of this report (Figures 1 and 2).

Information gathering for research and statistical monitoring

Key points

- There are gaps in statistical knowledge and research knowledge. Gender questions and analyses need to be routinely and systematically built into occupational safety and health monitoring and research.
- The closer the breakdown according to the real tasks people carry out, the better. Data should be adjusted for the number of hours worked.
- Practical means of improving the quality of European and Member State occupational safety and health statistics by gender should be investigated.
- The recommendations from the gender analyses of the third European survey on working conditions for modifications to the survey should be implemented.
- The gender imbalance in research should be addressed in all Member States, particularly as the situation and context in different Member States vary.
- More interdisciplinary approaches to gender research are needed.
- More research is needed in the area of workplace interventions and prevention.
- A European debate between researchers from different countries is needed to adopt a broader view.

Investigation of gender differences in occupational safety and health requires sex-differentiated statistics on working conditions and effective monitoring models for the complex work hazard situations encountered by both women and men.

In the introduction to this report, a note was made of some of the limitations of existing data from the EU for obtaining an accurate picture of gender differences in occupational safety and health exposures and outcomes. Nevertheless, the analysis that Europe's statistical office, Eurostat, has provided on Europe's accident and work-related health data, adjusted for hours worked, is very welcome (Dupré, 2002). Important, too, is the gender analysis of the third European survey on working conditions including its recommendations for improving the survey's gender-sensitivity.

Gaps in statistical information available from the Member States include not all countries providing information broken down by gender, and some Member States excluding some issues such as incidents of violence from members of the public.

The quality of gender statistics on occupational safety and health outcomes can be enhanced if they take account of the number of hours worked and division into occupations as well as division by sectors. The closer the breakdown according to the real tasks people carry out, the better.

Messing (1998) investigated gender deficiencies and bias in occupational safety and health investigations and information collection. She concludes that there are a number of steps that should be taken to improve the quality of available information:
- Gather information relevant to women's occupational health and analyse it in such a way that the health problems and their sources become visible.
- Improve the database for women's occupational health (job indications on death certificates and in governments and hospital records).
- Develop sensitivity to the implications of employment segregation for data collection (e.g. awareness that different health problems may result from different task assignments).
- Ensure use of data-gathering instruments that have been validated with and standardised for both sexes.
- Move towards recording accident statistics based on hours worked rather than on numbers of individual workers.
- Use measures complementary to accident statistics based on hours worked rather than on numbers of individual workers.
- Use measures complementary to accident statistics in order to identify risky jobs; include qualitative interview methods.
- Extend interest in women's reproductive health beyond foetal protection to include fertility, sexuality, early menopause, and menstrual disorders.
- Improve record keeping on absences due to sick leave.

Practical means of improving the quality of European and Member State occupational safety and health statistics by gender should be investigated. The important need of candidate countries to break down their statistics by gender has also been highlighted (Gonäs, 2002).

It is not possible to examine possible research bias in detail in this report. However, research is very often gender neutral, and gender may be included only when there is an individual researcher interested in such matters (Rantalaiho et al., 2002). Where both men and women have been included in studies, often the variable of sex is 'controlled' for, which means that gender differences cannot be examined. Two important issues are: the inclusion of areas relevant to both women and men in research programmes; and the gathering and analysis of data in a gender-sensitive way. If an area is neglected by research, then the risks remain unknown, which can perpetuate the belief that there is no risk, so the area is neglected both for prevention activities and for future research. A number of

Box 57: How much research attention is given to men's and women's work-related health?

Articles on various diseases and disorders discussed in this report were searched from the Medline (PubMed) database covering the years 1999–2002 and categorised according to their coverage of men and women. The results are shown in the following table:

Disease/disorder	Total	Word 'female' mentioned	Word 'male' mentioned	Female only	Male only
Accidents	62	37	46	1	10
Wounds and Injuries	281	173	189	13	29
Musculoskeletal disorders	109	79	76	11	8
Neoplasms	498	265	345	41	121
Lung diseases	267	141	175	7	41
Asthma	66	39	44	1	6
Skin diseases	132	88	93	4	9
Hearing disorders	14	9	14	0	5
Headache	2	2	2	0	0
Sick-building syndrome	7	6	6	0	0
Multiple chemical sensitivity	5	3	3	0	0
Infectious and viral diseases	343	197	171	50	24
HIV, hepatitis B, hepatitis C	154	103	99	19	15
Heart diseases	55	27	50	0	23
Mental disorders	125	97	94	11	8
Reproduction and infertility	214	173	95	93	15
Total	2 334	1 439	1 502	251	314

The search was conducted with the medical subject headings terms (MeSH) using major headings for the name of the disease linked with the sub-heading 'Epidemiology' and limited to human, gender female and/or male. Occupational dimension was searched with the MeSH terms 'occupational diseases' and 'occupational exposure', and with the text terms worker(s) and employee(s). There were no articles on common cold or tinnitus. On reproductive health, major headings were accompanied by menstruation disorders, pregnancy complications and infertility. Articles dealing with HIV, hepatitis B and hepatitis C cover a wide range of occupations from prostitutes to truck drivers and healthcare workers. Some healthcare workers with needle stick injuries, however, belong to the group of wounds and injuries.
'Total' refers to the number of all articles published on the topic in question. — it is not the sum of the rows, as the figures on each row merely reflect the search history.

Men were mentioned in more articles than women and more articles were found that mentioned only men than mentioned only women. There were more articles on neoplasms in all groups apart from 'Only women', where reproduction and infertility was the topic with the greatest number of articles.

specific areas for further research have been highlighted already in this report. There is a need for more gender-sensitive approaches to research to be developed and existing methodologies shared. Vogel (2002) also suggests that research suffers from policy compartmentalisation; for example, there is significant research on gender segregation in employment, but little of it addresses working conditions (see also Annex 11). There is a need for a more interdisciplinary approach. More research, in particular, is needed in the area of workplace interventions and prevention. A participatory intervention research method is presented in the section on involving women workers.

We have seen how gender segregation in jobs affects exposure to risks, so this must be taken into account in research study design. If men and women are exposed to different risks or different levels of risk, and this is not taken account of, then valid conclusions relating to gender differences seen in health outcomes cannot be made. Researchers such as Messing (1998) and Punnett and Herbert (2000) have stated that more studies of men and women (as far as possible) carrying out the same tasks in the workplace are needed. See also Messing et al. (2003).

The ILO has made suggestions on research to evaluate the real differences between the sexes and a report to the Global Commission on Women's Health — set up by a WHO resolution — makes proposals for improving research and data gathering (see Annex 8). In addition, NIOSH in the United States has put in place a specific research programme (see Box 58).

While there has been a greater amount of gender research coming from the Nordic countries (Vogel, 2002), research from the southern Member States has sometimes explored different fields and used different perspectives. A European debate between researchers from different countries would be useful to adopt a broader view on the different issues. The inclusion of experiences from research in the area of women's studies would also be useful. The gender imbalance in research should be addressed and research into gender and occupational safety and health issues in those Member States where there is less information available is necessary, particularly as the situation and context in different Member States vary.

Box 58
NIOSH gender research programme
NIOSH (the National Institute for Occupational Safety and Health in the United States) is carrying out specific activities on gender and occupational safety and health. Activities include an expanding research programme to address the occupational safety and health needs of working women. Research areas include: musculoskeletal disorders (for example, among women in the telecommunication, healthcare, service, and data entry industries); identification of workplace factors particularly stressful to women, and potential prevention measures; reproductive hazards; violence at work; women in non-traditional employment (e.g. in construction, including the use of tools, machinery and personal protective equipment); cancer (including possible work links to cervical and breast cancers); and the health and safety of healthcare workers.

Research projects in Italy
The Italian Institute of Occupational Health, ISPESL, is incorporating projects aimed at women workers into its programme research, for example:

- Multi-centric case control study in eight national birth centres on the relationship between occupational hazards and pregnancy.
- Study on the relationship between 'occupational exposure to lead' and 'post-menopausal osteoporosis'.
- Epidemiological studies on women workers in the clothing industry.
- Attitudes regarding safety at work and the level of mental well-being of the educational staff of young child services (zero to six years) in the municipality of Rome.

Source: ISPESL (2002).

Information and support for workplace activities

Key points

- Current scientific knowledge already provides enough information on occupational safety and health risks to women for preventive action to take place in a gender-sensitive way.
- While there is still a lack of information and support on gender and OSH for the workplace, some prevention services are beginning to integrate gender into their activities.
- More authoritative guidance, inspection tools and training on carrying out risk assessment in a gender-sensitive manner are needed.
- Existing examples of tools and programmes should be shared.
- Awareness raising, for example among OSH professionals and practitioners, is needed.
- Gender issues should be systematically integrated into all guides and support activities.

- OSH authorities, labour inspectorates and social partners have an important role to play.

Raising awareness and providing information, practical tools and instruments and technical support are important parts of supporting occupational safety and health risk-prevention activities in the workplace. There is evidence that targeted information and activities, aimed at specific jobs and tasks can be particularly successful (European Agency, 2001a, 2001b).

Sources of information include: journals for occupational safety and health practitioners; information and activities of occupational safety and health authorities, trade unions and employer associations; and joint guidelines produced though social dialogue.

The invisibility of gender in journal information

As has been mentioned, much of the research concerning occupational safety and health is gender neutral (see Niedhammer et al., 2000; Messing et al., 2003). Occupational safety and health journals are an important source of information for practitioners on prevention. To see how well gender issues are covered, a small review was conducted on Finnish and Dutch national journals and the results are presented in Boxes 59 and 60.

Box 59: Some characteristics of the national Finnish and Dutch OSH journals studied

English title	National title	Country	Time period	Number of articles
People and work	Työ ja ihminen	Finland	1997–2001	109
Journal for Occupational and Insurance Physicians	TBV (Tijdschrift voor bedrijfs- en verzekeringsartsen)	Netherlands	2000–01	89
Working conditions	Arbeidsomstandigheden	Netherlands	2000–01	216

NB: Only the English abstracts were used, not the original articles written in Finnish.

The articles were categorised according to how they analysed or presented the content or data: (1) men and women were dealt with separately; (2) only women were discussed; (3) only men were discussed; or (4) gender was not mentioned.

Box 60: The percentage of articles classified in regard to the analysis and reporting of the content and data in one Finnish and two Dutch OSH journals

Category	People and work	TBV	Working conditions
Men and women	11	2.3	0.9
Only women	1.8	3.4	0.5
Only men	7.4	0	0
Gender not mentioned	79.8	94.3	98.6
Total	100 (n = 109)	100 (n = 89)	100 (n = 216)

The vast majority of articles in all three journals did not look at gender. The gender-sensitivity in journals for practitioners may even be lower than in research journals.

Awareness-raising material, guides, practical instruments and technical support

It is beyond the scope of this report to analyse in detail the availability of occupational safety and health information and tools on a gender basis. Given that the traditional focus of occupational safety and health activity has been on accidents and on ill-health issues that are more present in male-dominated jobs, it is likely that there will also be more information and support available in these areas. Based on a survey, the European Trade Union Technical Bureau reports that prevention services are beginning to integrate the gender dimension into their activities, but generally only in 'female-dominated' sectors, or with regard to problems considered to be 'more specifically for women'. However, large country differences in approach were found, for example between Scandinavian and Latin countries (Vogel, 2002. See also Annex 11). The importance of the role of labour inspectorates for supporting improvements in women's working conditions has also been highlighted for candidate countries (Skiöld, 2002).

Nevertheless, targeted efforts to reach some sectors where women are typically employed could be especially important as these can be difficult sectors to reach — for example, service sector jobs such as cleaning, hotel and catering and private residential care for the elderly, and homeworkers.

Throughout this report, examples from the Member States of information material and practical support for the workplace have been presented (for example, see Box 61). Further development and sharing of awareness-raising material, practical guides and instruments and examples of good practice interventions targeted at female-dominated jobs, and the hazards to which women are particularly exposed are needed.

It is suggested that authoritative guidance, inspection tools and training on carrying out risk assessment in a gender-sensitive manner are needed. Attention to gender issues should be systematically integrated into all guides and support activities (see the section on gender mainstreaming and taking a gender-sensitive

---

Box 61

EU social dialogue results in practical guidance aimed at the jobs women do

One European initiative has been taken to redress the lack of information on preventing work hazards to cleaners, a predominantly female group who are low paid, and who often have part-time and precarious contracts. Within one of the European sectoral social dialogue forums, UNI-Europa, a European trade union federation and EFCI/FENI — European Federation of Cleaning Industries — developed a practical guide covering various risks to cleaners and their prevention.

Source: UNI-Europa and EFC/FENI (2001).

Portugal targets ceramics and textile sectors for campaigns

Following campaigns in construction and agriculture, Portugal's national OSH institute and labour inspectorate turned their attention to the textile and ceramic sectors, both high-employment areas for women. The resources for these campaigns include prevention instruments, training and events. Social dialogue is strongly promoted and incentives to carry out research and investigation provided. The involvement of employers' organisations and trade unions is a strong component in the development and implementation of the campaigns. The campaigns are also taken into schools and vocational training.

Sources: IDICT web site; European Agency (2001a); European Agency (2003c).

Italian awareness-raising activities

Italy's occupational health institute, ISPESL, has a national project financed by the Ministry of Health to increase awareness of occupational risks to women in various sectors and in relation to various risks.

Source: Papaleo (2002).

approach). As well as practical tools for the workplace, tools and training may be needed for labour inspectors and occupational health services. One study on the views of occupational physicians cited in the section on occupational health services underlines this point (Vinke et al., 1999).

Trade unions are another source of information and support and initiators of interventions (TUTB, 2000; Vogel, 2002). Among many guides available are Spanish trade union guidelines that cover risk assessment and prevention in the communications and transport sector (Comisiones Obreras, 2001) and a UK public sector union guide aimed at raising awareness and providing support to safety representatives in the workplace (Unison, 2001). Some examples of trade union instigated interventions are given in Annex 11. Another influential source of information and support could be joint employer–trade union guidelines, such as the example in Box 61.

Access to and functioning of occupational health services

Key points

- There is a lack of information regarding gender and use and functioning of occupational health services.
- There is some evidence that men may make more use of the services. Others have found no difference.
- There is some evidence that some practitioners may view male occupational health as being more important.
- Provision of occupational health services is highly dependent on the national context. It varies according to size of the organisation, with those working in small businesses less well covered, and will also be influenced by the higher priority given to 'men's' jobs and the risks found in them.

- Further research into gender issues in the provision and use of occupational health services is needed.

Occupational health services support good practices with regard to health, environment and safety management in organisations. They may provide risk-assessment and prevention services, health surveillance and monitoring, treatment and rehabilitation. Access to occupational safety and health services and where their resources are focused will have an influence on recognition and prevention of risks. Any gender differences in access to and prioritisation of resources may affect the health outcome of women compared to men.

According to a survey conducted by Hämäläinen et al. (2000), access to and provision of occupational health services vary greatly within the European countries. Some Member States have legislation establishing the State provision of occupational health services. In some countries, services are provided by insurance organisations, and some rely on provision through national health services. The extent to which different medical and technical disciplines are involved varies. Some organisations employ or contract nurses and doctors qualified in occupational safety and health and other specialists. However, small firm employees often have no access to occupational health services, partly because of cost factors to their firms (Graham, 2000). Again, service provision is likely to be on the basis of what are perceived to be the priority risks and priority sectors and who is considered to be most at risk.

Information on sex or gender differences in the frequency, amount or quality in traditional healthcare use is scarce. Very few studies have been conducted on access to occupational health services, and those that have, such as the study by Hämäläinen et al. (2000), do not look at gender. Also, research on the opinions and attitudes towards the occupational health services is scarce.

Studies have shown that women are more likely than men to visit their general medical practitioner and that, in general, medical consumption is higher among women (Verbrugge, 1985, 1986). This can also be related to their family responsibilities and because they live longer than men on average.

A recent longitudinal study in the Netherlands also showed that women were more likely to contact the general health practitioner than men, whereas men were more likely to contact the occupational health physician before and after a period of absenteeism (Houtman et al., 2002). This study suggests that women do not call upon medical assistance more frequently, but they differ from men with regard to whom they contact for medical advice. In contrast, in Finland, a large survey carried out in 2000 asked employed women and men about these issues and no differences were found: both sexes had used the services equally, they were equally satisfied, and appreciated equally the impartiality and confidentiality of the services (Piirainen et al., 2000).

The use of occupational healthcare may be a more complex phenomenon, and may also relate to health professions having different attitudes towards male and female workers. For example, occupational health physicians in another study in the Netherlands reported that it was less important for women to be rehabilitated into work than it was for men (Vinke et al., 1999). Although this study looked at intentions rather than actions, it may well be that these intentions resulted in sex-segregated action by, for example, occupational health physicians.

Another less researched question is the quality of relationships between patients and occupational health physicians. For example, according to *Eurobarometer 1996*, 36 % of European women prefer a female doctor. The profession of occupational health physician may be more male dominated than general medicine.

More research is needed to investigate the access and quality of services available on a gender basis and the reasons for any differences. We have seen above that the availability of services depends at least on country legislative differences and type of organisation as well as the attitude of the service provider.

Box 62

According to the European Women's Health Network, gender-sensitivity in occupational health services is necessary because occupational gender segregation is strong: men and women are exposed to different workplace environments and different types of demands and strains. It is important to acknowledge that certain health problems are unique to or have more serious implications for either women or men, and that they often have a link with age. Equal weight should be given to knowledge, values and experiences of both women and men, as full participation of both sexes in risk assessment and priority setting is necessary. The promotion of well-being and work satisfaction acknowledges also the importance of balancing work and private life for all employees. Mainstreaming gender into occupational health services emphasises the holistic view and significance of multiple exposures, both physical and psychological.

Source: EWHNET (2001).

Consultation and participation of women in workplace occupational safety and health matters

Key points

- Effective worker consultation and participation are a key factor in successful accident and ill-health prevention in the workplace. There is evidence that workplaces where there are trade unions have better safety records than non-unionised workplaces.
- Women have less power in the workplace as they are more likely to work in more junior or unskilled posts.
- Women are less likely to be members of trade unions, and are under-represented among union representatives compared to their overall membership levels.

- Women are less involved in occupational safety and health matters and decisions about these issues.
- There are examples of occupational safety and health intervention methods to involve women workers effectively. These examples should be shared.
- More models for the effective participation of women in occupational safety and health decisions at all levels are needed.

Women's weaker participation

Various studies by the European Agency for Safety and Health at Work have highlighted effective worker consultation and involvement as crucial factors for ensuring effective preventive interventions in the workplace and management of occupational risks (for example, European Agency, 2001a, 2002b). Other research has highlighted the importance of women's presence in collective bargaining for equality (Dickens, 1998). Therefore, the participation of women workers as well as men in occupational safety and health matters is also crucial if risks to both women and men are to be tackled effectively.

Respondents to the European Foundation's third survey on working conditions were asked to comment on the possibility they had to discuss working conditions and organisational change. The analysis suggests that exchanges on working conditions and organisational change are more frequent among those in qualified occupations and with a permanent job. Unskilled workers are the least involved in these exchanges and temporary agency workers have the least opportunity to discuss either working conditions or organisational change. The survey also looked at who takes part in these exchanges and found that male workers are more involved in exchanges that involve staff representatives and outside experts than women. It seems, then, that women are less involved than men in exchanges where more detailed discussion and decisions about safety and health in the workplace are likely to take place (Paoli and Merllié, 2001). A Greek study of women's participation in occupational safety and health at the workplace found that, compared to men, women were more likely to take a passive role, they were less involved and they had a high level of distrust of the OSH management process (Batra, 2002).

One important form of consultation and participation in workplace health and safety issues is through trade union representation. Various studies report that unionised workplaces carry out more safety management activities and have better safety records (for example, Reilly et al., 1995; Steele et al., 1997). However, a lower percentage of women workers are members of trade unions than male workers and their presence is even lower among safety representatives and worker representatives. For example, membership of trade unions affiliated to the Trades Union Congress in the UK is 59 % male and 41 % female. However, of respondents to a survey of workplace safety representatives from these trade unions, 72 % were male and 28 % were female (TUC, forthcoming). For a discussion of women's weak representation within trade unions relative to their membership, see Garcia and Lega (1999).

Rosskam, reporting experiences from her involvement in the ILO 'Safe work' programme, points out that women are absent as decision-makers at all levels (see also Messing, 1999), as there are certain recognised obstacles for women taking on leadership roles and, in addition, for occupational safety and health, they may be discouraged by the perceived technical nature of the topic. As a result, there is a lack of participation of women in safety committees, which exacerbates the lack of participation by women in solutions affecting their own health. It is proposed that trade unions should be encouraged to 'activate women's committees as well as involve women in health and safety committees ...' as part of the solution to this problem. A similar recommendation is made in an action plan on women's health in the workforce, drawn up jointly by researchers and trade unions in Quebec, Canada (Cinbiose web site). Some trade unions have produced guidance for the workplace on encouraging and facilitating more women to become safety representatives (for example, see Unison, 2001, and Box 63).

Box 63

Workplace safety representatives can raise awareness of, investigate and put women's health and safety on the agenda by:

- raising the health and safety problems women face;
- encouraging more women to become safety representatives;
- mounting health and safety campaigns aimed at women;
- encouraging women to attend training courses;
- finding other ways of raising awareness of women's health and safety concerns;
- ensuring there are safety representatives appointed in areas where mostly women work;
- organising meetings when part-timers and women with childcare responsibilities can attend;
- providing information about women's health and safety concerns;
- carrying out a simple survey to find out more information about women's health and safety concerns and talking to women during inspections;
- finding ways to keep in touch with women who work part-time or unsociable hours (e.g. cleaners).

Source: Unison (2001).

Improving women's participation is also among the recommendations in a paper published by the ILO on integrating the gender perspective into the field of occupational safety and health: 'Women should be better represented and more directly involved in the decision-making process concerning the protection of their health. Women's views as users, care givers and workers, and their own experiences, knowledge and skills should be reflected in formulating and implementing health-promotion strategies. They should have a greater participation in the improvement of their working conditions, particularly through programme development, provision of occupational health services, access to more and better information, training and health education. The support of women workers to organise themselves and participate in the improvement of their working conditions should be reinforced at the national and enterprise level' (Forastieri, 2000).

A stepwise, participatory approach to prevention

A stepwise, participatory approach has proven useful in prevention, including in relation to the prevention of stress and musculoskeletal disorders (for example, Kompier and Cooper, 1999; Kompier et al., 1996; Landsbergis et al., 1999). A typical example of this type of approach is that developed by Kompier and Cooper (1999):

1. *Preparation:* What is the problem? Is all information available to tackle the problem or should more information be gathered? Who plays a role in the preventive action and what is their attitude towards prevention? Both management and employees have to support the action, particularly to implement organisational measures.

2. *Problem analysis:* In complex situations, additional collection of information is necessary. Instruments, often a combination of different tools, are used to obtain all relevant information. In organisational settings, the identification of risk groups is particularly important.

3. *Choice of measures:* Both work-directed and person-directed measures can be decided

upon, depending on the risk groups and risk factors identified. For example, for prevention of musculoskeletal disorders, work-directed measures included work redesign, changes with respect to work and rest-time policies, social support, and ergonomic and technological actions. The most important person-directed actions relate to the training of the employees and management, or providing personal measures, such as coaching. It is important to understand that some benefits can be expected in the short term and some in the long term.

4. *Implementation:* Measures have to be implemented. Although taking action may seem self-evident, this step is often the most problematic one: case studies show that it is precisely here that the risk of failure is greater.

5. *Evaluation:* This is the final step, which directs future action. It is very important to evaluate both the process (Did everything go according to plan? Where did things go wrong? etc.) and the outcome, for example the reduction in lost days/absence rates, etc.

Box 64

In the Netherlands, a hospital was eager to reduce sickness absence. The hospital also wished to increase its competitiveness and attract more personnel. The management took the initiative and a steering committee was formed. Employees from different levels and departments in the organisation were asked to join the committee. This improved the degree of commitment and democracy and ensured the flow of information. In addition, an external consultant, and energiser, was hired to provide advice and keep the process going. There was an improvement in working conditions, improved patient care, and a better psychosocial climate. Absence percentage decreased from 8.9 to 5.8 %. However, it appeared to be difficult to keep the middle management committed. The members of the committee kept changing, the responsibilities of the steering committee were not clear to all, it was difficult to keep the employees involved in a four-year project and it was difficult to assess objectively which were serious constraints and which were not.

Source: Kompier and Cooper (1999).

Some methods for improving or assisting women's participation within this approach are given below.

Examples of methods and approaches to facilitate women's participation

There are various initiatives, methods and approaches that have been adopted and recommended to facilitate a greater involvement of women workers in the process of researching and assessing workplace risks and introducing preventive measures. These include the following:

- *Action-oriented research*. This is a method of intervention research to ensure effective participation of the workers involved when studying problems and proposing areas for improvement, and to ensure that the real tasks being done by women and men are assessed (see Box 66).

- *Health circles and small discussion groups*. Getting women workers together at the workplace to discuss their work-related health problems is one part of promoting women's participation. One example of how to do this is the Spagat project, from Austria. This is a method for involving working women, using their experience and knowledge, in making a systematic analysis of their health problems, and developing practicable solutions. It involves the cooperation of the company management, trade union committee and workers, and uses 'health circles'. The health circles are working groups set up within an enterprise, where small groups of employees use a series of meetings to analyse unhealthy situations at work, and develop ideas for workable solutions. The project was developed by a research and training organisation, in cooperation with a trade union federation and with support from the Austrian Health Fund FGÖ (Spagat project web site).

General advice on running small discussion groups in the workplace on safety and health is given in a 'Barefoot research' manual on worker involvement in health and safety issues developed for the ILO (Keith et al., 2002).

Box 65

In Germany, prevention and health promotion have taken the form of 'health circles'. This is a method that can be described as follows (e.g. Sochert, 1999; Mather and Peterken, 1999):

- a small group of employees and their management within an occupational or sectoral entity is formed;
- the group comes together in several meetings over a certain (agreed upon) time period;
- an independent moderator leads these sessions;
- during these meetings, the demands are discussed in relation to health, and suggestions to improve health for the employees are discussed.

- *Hazard 'mapping' research methods.* Mapping is a participatory method for surveying workers for work-related health problems described in the ILO 'Barefoot research' manual (Keith et al., 2002). Using simple outline drawings of the body or plans of the workplace, workers can record, in a visual form, their health problems, work hazards and overall work environment. It can work well, for example, to investigate symptoms of more complex health issues such as musculoskeletal disorders. It works best among a group of workers carrying out the same task. A group of workers, with the help of a facilitator, each mark their aches and pains on the body map. Clusters of symptoms that appear on the same part of the body are likely to be work-related. Isolated symptoms are less likely to be work-related. In this way, it is used to identify common patterns of health problems amongst workers in a particular workplace or doing the same job. As it is done in groups, it is participatory and develops a collective approach. It can be more easily understood and more inviting than filling in, on an individual basis, a workplace survey form. It has been used by trade unions with workers to investigate allergies, musculoskeletal disorders, reproductive hazards and stress.

- *Trade union initiatives.* Many trade unions have produced specific guides, run campaigns and organised specific training regarding women and occupational safety and health risks, aimed at improving awareness of gender and occupational safety and health issues and involving women workers. Some examples have already been given. The 'Barefoot research' manual published by the ILO gives various practical examples of workplace inspections and interventions covering women's tasks and involving women workers (Keith et al., 2002). Some additional examples of trade union initiatives can be seen in Annex 11.

- *Developing cooperation and common plans between researchers and representatives of women workers.* An example of this is an action plan, developed by a meeting of Canadian researchers and representatives of women workers in 1998 at the Université du Québec à Montréal (Cinbiose web site, referred to above). The action plan defines a number of issues relating to women's health at work, and in each area sets goals and proposes means for achieving the goals. The plan addresses legislation and policy, and practical measures as well as research areas.

While more models for the effective participation of women in occupational safety and health decisions are needed, the methods and examples of their application that already exist should be exchanged and shared.

Box 66: Example of worker participation through action-oriented research interventions

Worker participation is a key factor in the success of health and safety interventions. However, because of their concentration in lower status jobs and because of their lower representation within trade unions, women have less access to and involvement in decision-making processes. Therefore, for interventions aimed at female-dominated work groups, effective participation of women workers themselves is particularly important. With this in mind, the Cinbiose centre at the University of Quebec has developed an action-oriented research approach to workplace interventions that it has used to look at ergonomic problems in predominantly female work groups. The principles behind the approach are:

- empowerment;
- using workers' knowledge;
- questioning established assumptions about risks at work;
- recognition of gender-based issues in the division of labour and power relations.

The aim is to incorporate knowledge of men and women workers into all stages of the interventions. On-the-spot observation of those doing the work and questioning of this group, as well as their supervisors and trade union representatives, are used to suggest changes based on real work and set priorities.

Methods used include:

- preliminary study using group interviews and collecting background information such as accident records to help target jobs for study and set priorities;
- preliminary observation of the selected jobs to target further particular activities to be studied;
- systematic observation and analyses of the real work activities, including the collection of quantitative and qualitative measurements, observation of workers and using questionnaires and interviews with the workers concerned;
- feedback and validation, by presenting and discussing the results with workers, supervisors, union representatives, safety committees, etc.;
- dissemination of results.

An example of the application of this methodology is a study of the clerical work of receptionists dealing with admissions in a hospital. The work is typically considered as low risk and physically undemanding: comfortable surroundings; little decision-making. Problems uncovered in the study included:

- communication and information management — working over the telephone in a noisy and busy environment, and locating, establishing and clarifying required information with patients, within their records or using the knowledge of other receptionists;
- frequent interruptions of tasks, adding to the mental load of the task;
- lower back pain and circulatory problems from constant sitting; pain in neck and shoulder area, where causes included simultaneous telephone use and record retrieval activities;
- workstation layout problems, for example: work surface of insufficient depth and too high; computer screens too high; walking distances between work equipment and workstation and to call patients.

The report resulted in a number of recommendations, such as the use of cordless phones, better information signs for patients so they do not need to ask receptionists, improved work layout and a patient call system and proposals to redesign the plan of the offices, acquire earphones and update the computing system.

Source: Messing (1999).

6. GENDER MAINSTREAMING

Occupational safety and health legislation

Key points

- Traditionally, occupational safety and health legislation was often discriminatory, banning women from performing certain work and concentrating on risks in male-dominated occupations.
- Modern legislation takes a gender-neutral approach.
- Although the general EU legislation on prevention covers health problems and multi-causal risks which are more often a feature of women's work, such as stress and upper limb disorders, hazards in male-dominated work are still better covered by specific legislation.
- There are some exclusions to the directives that impact more on women, such as the exclusion of paid domestic workers.
- Although EU directives apply to the public sector, where women are over-represented, restricted enforcement arrangements operate in some Member States.
- It is possible to apply the gender-neutral risk-assessment and prevention approach in a gender-sensitive way. Guidelines, risk-assessment tools and training are needed to support this. In addition, gender assessments should be carried out on directives when they are reviewed or new ones planned.
- If the EU directives were effectively implemented and enforced in all EU workplaces, this could mean a significant improvement in health and safety at work for many women.
- Women's working conditions need to be better promoted in the standardisation process.
- The effective implementation of employment equality measures would also contribute to the removal of some sources of stress in women's work.

The historical approach to occupational safety and health and gender

Historically, occupational safety and health legislation was focused on safety, accidents, machinery hazards and high-risk occupations such as mining. It has already been discussed that men suffer more accidents at work and that women are more affected by certain occupational diseases. Another approach historically was to place restrictions on the employment of women in certain occupations or circumstances. This approach, often based on very limited scientific information, had discriminatory consequences for women's employment rather than improving the working conditions for both genders. For example, at the beginning of the 20th century, women were banned from working with lead. They were more exposed because of the jobs they did, but there is no great gender difference in the ill-health effects of lead. The result was that men were left working unprotected in the same hazardous conditions.

Current EU occupational safety and health legislative approach

EU occupational safety and health directives have abandoned this discriminatory approach. Current legislation on occupational risk prevention generally takes a neutral approach with regard to gender.

The 1989 EU framework directive (Council Directive 89/391/EEC) sets the general approach for current risk prevention in the EU. It is a general directive, requiring employers to identify and evaluate occupational risks, without limiting the scope of what those risks may be, and to bring in adequate protective and preventive measures, based on a hierarchy of measures, starting with prevention at source. The approach is that work, workplaces and work equipment should be adapted to the safety and comfort of the worker. Psychosocial issues such as stress, violence from members of the public and work organisation are covered by the general provisions of this directive, as well as work-related limb disorders, although they are not covered by specific directives. On the other hand, there are now specific directives on chemicals, biological agents and manual handling activities, for example. The hierarchy of preventive measures given in Article 6, starting with avoiding risks and giving collective measures precedent over individual measures, includes adapting work to the individual with respect to workplaces, work equipment, working methods and work organisation. However, the framework directive specifically excludes paid domestic workers, a job dominated by women. In addition, it should be remembered that the historical past of legislation, focusing on men and the jobs they do, means that the development of today's 'gender-neutral' legislation has been based on male norms.

Women are over-represented in the public sector, to which the EU directives also apply. However, the directives do not cover methods or standards of enforcement, and in some Member States, in some parts of the public sector, different, more limited provisions for labour inspection and enforcement operate and this can have a negative impact on the implementation of health and safety standards in these workplaces, including use of prevention services, training and worker representation (Vogel, 1999; see also Maestro et al., 2000; Montero and Frutos, 2000).

As discussed in the section on reproductive hazards, recognising the special childbearing capacity of women, the EU has also introduced specific legislation covering the safety and health at work of pregnant workers and new mothers. Council Directive 92/85/EEC places requirements on employers to assess the risks to these workers and take measures to avoid the risks, restrictions on night work and employment rights issues. However, the European Trade Union Confederation (ETUC) is among those that believe that this legislation is not sufficient, for example, because it allows for the removal of pregnant women from hazardous work, rather than putting more emphasis on prevention at source through collective measures (Vogel, 2002). Also, there is no specific legislation on broader aspects of reproductive hazards, which would cover all aspects of both men's and women's reproductive health.

Although occupational safety and health directives are based on a general risk-assessment and prevention approach, it is possible to apply them in practice in the workplace in a more gender-sensitive way. Box 67 highlights features of some of the directives that could facilitate taking a gender-sensitive approach.

Some issues relevant in particular to women's occupational safety and health are covered by legislation outside the occupational safety and health framework. Council Directive 76/207/EEC, as amended by Council Directive 2002/73/EC of 23 September 2002, dealing with equal treatment for men and women as regards access to employment, vocational training and promotion, and working conditions, covers sexual harassment and additional measures relating to pregnant women.

With their broad approach, if the EU occupational safety and health directives were effectively implemented and enforced in all EU workplaces, this could mean a significant improvement in health and safety at work for many women.

> **Box 67: Some examples of gender-neutral occupational safety and health directives and elements that could be applied in a gender-sensitive way**
>
> The framework directive can generally be said to take a gender-neutral approach in its aim to provide for minimum standards of safety and health. It requires risk assessment and prevention and covers musculoskeletal disorders and psychosocial hazards and outcomes, as well as the 'traditional' hazards. Its broad approach allows it to be applied in a 'gender-sensitive' way. In addition, Article 16 states 'particularly sensitive risk groups must be protected against the dangers which specifically affect them'. It is not specified what these 'sensitive groups' could be, so the article could be applied to gender differences, or issues such as older workers or workers with disabilities.
>
> The directive on work equipment (Council Directive 89/655/EEC) includes the general obligation that work equipment should not be prejudicial to the safety and health of workers and a provision that ergonomic principles must be taken into account.
>
> The directive on personal protective equipment says that the selection of personal protective equipment must take account of personal parameters relating to the user and the nature of his/her work.
>
> The workplace directive — Council Directive 89/654/EEC — concerns the minimum safety and health requirements for the workplace such as providing suitable workstations. It also provides some additional requirements concerning pregnant women. Minimum requirements specified in annexes include: 'Pregnant women and nursing mothers must be able to lie down to rest in appropriate conditions'; and 'in rest rooms and rest areas, appropriate measures must be introduced for the protection of non-smokers against the discomfort caused by tobacco smoke'.

Gender equality legislation and actions

Gender equality in general, at work and outside, is largely outside the scope of this report. However, we have seen that gender differences in status and treatment and in access to resources, both in society and in the workplace, can have an influence on gender differences seen in occupational safety and health outcomes. For example, discrimination in pay and promotion and differences in access to training can be sources of stress. The EU has introduced various directives, agreements and acts aimed at achieving equal treatment (see Annex 12). Again, if these were fully implemented in workplaces, this would also contribute to an improvement in women's health at work. In addition, some proposals for actions and strategies in this area, based on research of the European Foundation, are given in Annex 14. It could be particularly fruitful to try to obtain closer links between health policies at work and gender policies (for example, see Vogel, 2002 and Annex 11). This is discussed in the section on mainstreaming.

Standard setting and compensation

As has been discussed in a previous section, compared to preventive legislation, compensation coverage is much more confined to traditional risks found in many men's jobs. Many European and international standards for equipment or exposures were developed mostly on male data (for example, see Messing, 1998). Women's working conditions should be covered better in standardisation work.

EU strategy on occupational safety and health and gender

Key points

- The Community strategy 2002–06 includes mainstreaming the gender dimension into occupational safety and health.
- The European Parliament has highlighted some specific areas of attention regarding occupational safety and health and gender equality.

As was stated in the introduction, the European Commission has decided that there is a need to consider more closely the occupational health and safety of women and has highlighted the gender issue in its latest strategy for safety and health at work. It notes the increasing participation of women in employment, and acknowledges gender differences in the incidence of occupational accidents and diseases. The strategy puts forward a number of objectives for a holistic approach to well-being at work that must be targeted jointly by all players. These include 'mainstreaming the gender dimension into risk evaluation, preventive measures and compensation arrangements, so as to take account of the specific characteristics of women in terms of health and safety at work'. The European Parliament is invited to discuss and formally put forward its views on the Community strategies. In its comments on the occupational safety and health strategy, it strongly emphasised the importance of the gender element and proposed a number of ways of taking it forward.

Elements of the Community strategy on occupational safety and health 2002–06 relevant to gender and the improvement of women's working conditions include:

- gender mainstreaming as one of a number of objectives to achieve a holistic approach to well-being at work;
- mainstreaming the gender dimension into risk evaluation, preventive measures and compensation arrangements;
- gender mainstreaming to be included in the setting of national quantified objectives for reducing occupational accidents and illnesses;
- the intention to look at legislative and non-legislative approaches to musculoskeletal disorders, ergonomics, work-related stress, violence at work and psychological harassment at work;
- unpaid family helpers as the target of specific measures in terms of information, awareness and risk-prevention programmes;
- analyses of new and emerging risks, taking account of changes in forms of employment and work organisation.

The European Parliament comments on the gender dimension of the Community strategy included highlighting the following areas for attention:

- incorporating gender mainstreaming throughout the strategy, for example in relation to temporary and part-time workers;
- the 'double workload' of women in paid employment with household and caring responsibilities at home;
- differences in vocational education and training and occupational safety and health aspects that hinder women entering certain sectors;
- homeworkers and women in agricultural and small, family-run businesses;
- gender to be taken into account in guides and training materials, and future legislation;
- considering whether gender-specific needs have been taken sufficiently into account in legislation, even in sectors where women are under-represented.

Further details of the Community strategy and the European Parliament's position with regard to gender are given in Annexes 3 and 4. Information about approaches to gender issues in occupational safety and health in the international arena are given in a later section.

Mainstreaming, gender and occupational safety and health

Key points

- Mainstreaming gender into occupational safety and health is considered an important part of achieving improvements in occupational safety and health by the European Commission and by international occupational safety and health organisations.
- Gender assessments are an important tool in the mainstreaming strategy and there are a number of occupational safety and health areas where they should be carried out. Their use is recommended, given the gender neutrality of the current occupational safety and health approach.

Throughout this report, it has been seen that gender differences in occupational safety and health are influenced by a very broad range of factors both inside and outside the workplace. Yet, health at work, equality and public and social health policies are often very compartmentalised (Vogel, 2002. See also Annex 11). 'Mainstreaming' is the approach being used to promote the joining-up of policy areas.

The European Commission strategy on promoting equality between women and men in Commission policies and actions is based on 'mainstreaming'. In the section above, we have seen that gender mainstreaming has been incorporated into the EU strategy on improving occupational safety and health in Europe. Mainstreaming gender into occupational safety and health means actively integrating gender issues into information gathering, standard setting, legislation, preventive actions and policy setting and involving women in decision-making bodies, such as national safety councils, occupational health services, and enterprise-level safety committees. We have seen that, in its strategy on occupational safety and health, the Commission has adopted the goal of mainstreaming gender into its actions. Other bodies such as the WHO and ILO are also advocating this approach (see Annex 8) and the strategy of NIOSH to mainstream gender into its research programme has been mentioned.

'Gender impact assessments' are an important tool in the Commission's gender mainstreaming strategy and it has developed a general guide on them (European Commission, no date). As the Commission states in the guide, 'policy decisions that appear gender neutral may have a differential impact on women and men, even when such an effect was neither intended nor envisaged. Gender impact assessment is carried out to avoid unintended negative consequences and improve quality and efficiency of policies.' The guide explains that 'policies which appear gender neutral may, on closer investigation, turn out to affect women and men differently. Why? Because we find substantial differences in the lives of women and men in most policy fields.'

It is intended that this general guide should be adapted to the specific needs of each Commission policy area, as appropriate, and it is logical that this should be done for the Commission's policy work on occupational safety and health, given that a gender-neutral approach has also been taken in this area.

When European safety and health at work directives are being planned or existing directives assessed or reviewed, a gender impact assessment should be carried out as part of the process. Other areas for gender impact assessment in the occupational safety and health field include the setting of standards and exposure limits, compensation arrangements, setting benchmarks and targets for improvements in occupational safety and health, and priority setting.

Looking at the gender impact of benchmarking and target setting is important. For example, if objectives are set focusing particularly on the 'traditional' risks in male-dominated jobs or on types of accidents that occur more to men, then the gender impact will be uneven.

As well as applying 'gender equality mainstreaming' to OSH policy and actions — i.e. integrating gender issues into all occupational safety and health issues — OSH should be 'mainstreamed' or integrated into policy and actions on equality between women and men, such as employment and health equality issues. This is discussed in more detail in the next section.

Below are some suggestions for action areas for mainstreaming gender into occupational safety and health:

- improve gender-sensitive indicators for occupational safety and health and work organisation;
- include gender issues as appropriate in all Commission occupational safety and health guidelines and information activities;
- include looking at gender issues in the work of all ad hoc groups set up to advise the Commission on occupational safety and health matters;
- include gender in risk-assessment activities.

In some countries, gender mainstreaming in occupational health is already becoming standard practice. For example, in Finland, a gender impact analysis was made during the preparation of the Occupational Safety and Health Act (738/2002), which entered into force at the beginning of 2003. This was the first time that such a gender impact assessment had been made and it was justified on the grounds that working life in Finland is strongly gender segregated, which means different working conditions and risks for women and men.

The 'Worklife' and EU enlargement project is encouraging and where necessary supporting candidate countries in carrying out activity plans. For example, the plan from Malta includes development of a training manual on gender and equality guidance (Skiöld, 2002). Encouragement of such activities both within the Member States as well as the candidate countries is important, as will be sharing the results.

Box 68

Diversity among women and men needs to be acknowledged in policies, strategies and approaches. Mainstreaming gender equity in health is a strategy that promotes the integration of gender concerns into the formulation, monitoring and analysis of policies, programmes and projects, with the objective of ensuring that women and men achieve the best possible health status. Mainstreaming gender in health is both a political and technical process, which requires shifts in organisational cultures and ways of thinking, as well as in the goals, structures and resource allocation. A mainstreaming strategy does not preclude initiatives specifically directed towards either women or men when needed. Positive initiatives are necessary and complementary in a mainstreaming strategy (WHO, 2002a).

Mainstreaming occupational safety and health into equality initiatives

Key points

- National and workplace equality plans can be used to promote equality in working conditions.
- Gender-sensitive risk assessment can be used as a tool for helping to reduce inequalities in pay and opportunities.
- There are already examples of inclusion of occupational safety and health in equality plans. More research is needed in this area

- and existing examples of practice should be shared.
- Occupational safety and health should also be included in health equality programmes.
- Equality, including working conditions, should form part of corporate social responsibility (CSR) activities.

Achieving equality in occupational safety and health should also be part of the goals of employment equality policy, programmes and activities.

Gender-sensitive risk assessment, looking at the real jobs and tasks carried out, often shows that women's work is no less physically demanding than men's and often involves additional physical strains as well as pressures and stressors that are less common in men's work (Messing, 1998, 1999). If gender-sensitive, risk assessments of tasks and job responsibilities can be tools for helping to reduce inequalities in pay and opportunities (Grönkvist and Lagerlöf, 1999). Issues such as sexual harassment clearly relate to both equality and occupational safety and health.

Some Member States have legislation that requires employers to introduce equality plans. These equality plans can promote equality and working conditions if they cover job differentiation and health and safety risks (Grönkvist and Lagerlöf, 1999), and examples of their use in this way are beginning to appear.

Kauppinen and Otala (1999) carried out a project called 'Gender equality, work organisation and well-being' that aimed to define equality standards for a good workplace. The initiative came from the national labour-market organisation and the project was carried out in close collaboration with the employers' and employees' central labour organisations. In the 10 participating workplaces, women employees noticed more shortcomings in the equality atmosphere than men: 77 % of men but only 42 % of women said that gender equality was implemented well at their workplace. The project resulted in the development of eight equality standards for use by enterprises to make a self-assessment of their performance, that cover: the level of equality; salary and remuneration policy; career and work opportunities; common goals and opportunities for influence and control; work atmosphere and feeling of togetherness; information flow and openness in information delivery; working conditions; and reconciling work and family life (see Annex 13). The standards can be utilised in the planning of equality guidelines.

Box 69: Valmet Automotive — excellence in equality

The Valmet Automotive car manufacturing plant (Finland) introduced an internal equality programme in 2002. The programme's goal was to integrate gender equality and good working conditions. Equality in Valmet Automotive means that in everyday working life everyone, regardless of age, gender, ethnic background or position, is treated with respect. The equality programme includes the aims of providing every worker with safe and sound working conditions, taking into consideration aspects ranging from ergonomic work to combining work and family. The programme explicitly prohibits sexual harassment at the workplace. Active development of team-working skills aims to increase the employees' ability and chances to take responsibility for the development of their own well-being both as individuals and as a team. The equality plan will be taken into account in developmental projects, and plans for occupational health services.

Source: Ahjo nro (2002).

The EU sets employment guidelines for the Member States to follow. For example, 2002 strategy areas included increasing women's participation in the workplace, and issues for Member State consideration included gender equality, health and safety at work, flexibility and security, inclusion and access to the labour market, work organisation and work–life balance. Member States translate the guidelines into national policies and national action plans to develop the strategy for employment and should adopt a gender mainstreaming approach (European Council, 2002). As with workplace equality action plans, national plans in the area of equality in employment can promote equality in working conditions if they include occupational safety and health.

More attention is now being paid to achieving equality in health programmes and policies. Occupational safety and health should also be included within these programmes. Equality in working conditions should also be incorporated into corporate social responsibility (CSR) activities.

Gender equality in and outside the workplace

Key points

- Measures to improve equality would improve the quality of women's work.

- Other European bodies have proposed actions in this area — see Annex 14.

It was recognised at the beginning of this report that a whole range of issues relating to equality in employment as well as in society have an impact on gender differences in occupational safety and health in the workplace itself. Some issues such as the greater burden of work carried out by women in the home, sexual harassment and discrimination of women in the workplace have been mentioned in this report. For example, commenting on findings of the European survey on working conditions, Fagan and Burchell (2002) state 'some gender differences are due to the broader pattern of gender relations and inequality in society that transcend the focus on gender-segregated employment conditions'. Investigating broader equality issues is largely outside the scope of this report, and others at the EU level have put forward strategies for action in this area. Details of proposals for actions to reduce inequalities and examples of good practices based on the research of the European Foundation (2002c) are given in Annex 14.

Box 70

In the European Foundation report *Promoting gender equality in the workplace*, Olgiati and Shapiro (2002) (see also European Foundation, 2002c) discuss the use of equality plans in the workplace by investigating 21 case studies in seven countries. The results suggest that both the legislation and national programmes play important roles in initiating, sustaining and raising the profile of equality action in the workplace. Collective bargaining is also an important part of this process. Despite the importance of legislation, the results may be modest if the basic attitude of the company is negative. According to the report, most employers welcomed women's greater participation in the labour market, but pressure on companies to provide a 24-hour/seven-days-a-week service has a negative impact on balancing family and work.

Olgiati and Shapiro (2002) refer to the importance of equity at work. Equity refers to organisational justice and fairness of treatment in terms of rights, benefits, obligations and opportunities. It means equal opportunities for women and men and ability to make choices without limitations set by strict gender roles. Equity does not mean that all employees should be treated the same. For example, training and skills development should be adjusted to the employee's age, background and education. Authors such as Kivimäki et al. (2003) provide evidence that equity at work has a positive effect on employees' job satisfaction and mental health.

7.

DISCUSSION AND CONCLUSIONS

Improving the employability of women and men includes ensuring that they do not have to leave the workforce through injury or ill health, ensuring work is compatible with home life and ensuring that both women and men can work safely and healthily in different areas of work. There is also a high economic cost of failure to adequately prevent risks to both women and men at work.

We have seen that there are gender differences in the many factors that influence working conditions and quality of work. The factors are often interrelated. They include: legislation and social policy; perceptions and assumptions about who is at risk and what risks are found in different jobs, and what are included as work-related risks; discrimination both within and outside the workplace, gender relations and gender differences in society at large.

There are two important issues within this: factors that influence the type of hazards arising from work in the first place, and then factors which affect the extent to which these hazards are recognised and risks prevented.

Among factors that influence gender differences in exposure to hazards arising from work in the first place, and therefore differences in the occupational health outcomes, there are two critical ones: firstly, gender segregation in employment (job) and, secondly, gender-segregated domestic responsibilities together with the work–life interface.

Gender segregation at work

Because of strong occupational gender segregation in the EU labour market, which remains high despite changes in the world of work, women and men are exposed to different workplace environments and different types of demands and strains even when they are employed by the same sector and ply the same trade. There is strong segregation between sectors, between jobs in the same sector, and, as we have seen, there can be segregation of tasks even when women and men have the same job title in the same workplace. There is also strong vertical segregation within workplaces, with men more likely to be employed in more senior positions. For all the different types of hazards, both physical and psychosocial, job segregation strongly contributes to work exposure to hazards and therefore to health outcomes. In general, men suffer more accidents and injuries at work than women, whereas women report more health problems such as upper limb disorders and stress.

Other gender differences in employment conditions have an impact on occupational safety and health: more women are concentrated in low-paid, precarious work and this affects their working conditions and the risks they are exposed to; gender inequality both inside and outside the workplace can have an effect on women's occupational safety and health and there are important links between wider discrimination issues and health; more women work part-time, and part-time and full-time jobs have different characteristics; and men are more likely to work very long full-time hours.

Gender segregation at home

Unequal sharing of housework and the fact that women are more likely to have additional family and carer responsibilities put extra pressure on many women workers. This also increases the cost to the economy of women's work-related illness. A conflict between work and home is a source of work-related stress. Furthermore, it is important to acknowledge that the long and often inflexible working hours of men can also affect women's employment and family roles.

Gender-sensitive assessment — assessing the real exposure of women workers

Risks to women at work may be overlooked in various areas. Factors influencing occupational safety and health outcomes are many and often interrelated. For all the hazards examined in this report, possible reasons why the known risks to women may be underestimated were discussed. There are at least two important reasons for underestimation: first, if a holistic view of the sources of OSH risks and health outcomes is not taken, certain risks and health-related outcomes are not taken into account. Second, if false assumptions are made about who is at risk, some groups are erroneously excluded from the picture.

If women or the jobs they do are ignored in research, then risks are overlooked. This in turn leads to them being ignored for both prevention and future research. Again, if statistical monitoring does not look at gender issues, this distorts the picture. Good information is needed both for successful prevention and for effective policy-making and strategy formulation. OSH policies, legislation, prevention programmes, provision of occupational health services and actions by labour inspectors are based on risk analyses. If risks to women are hidden, this will result in unequal treatment.

Regarding gender segregation, we saw how even women and men with the same job description can carry out different tasks. This means that in research, statistical monitoring and interventions there is a need to look as closely as possible at the real tasks people carry out.

To get a true picture of the risks to both women and men, it is important to take account of the amount of exposure. This means adjusting for both part-time and full-time work, and the fact that in some sectors longer hours are worked. Where this is taken account of, the difference, for example, in accident or hearing disorder rates between women and men is reduced, although not eliminated. However, 'less at risk' is not the same as 'not at risk', although it appears that often it is interpreted as such. This can be the reason why women or gender are ignored in research into work-related health outcomes more prevalent in men, or sectors where women work are ignored. It can result in women in female-dominated jobs being excluded from workplace health checks, physicians failing to ask questions about the work-relatedness of women's health problems or women or their jobs receiving fewer resources from occupational health services.

Men suffer more discrete accidents, or health problems with a clear occupational cause such as hearing loss that arise from more visible risks and that have more obvious methods for their prevention. Outcomes such as upper limb disorders and stress, arising from the jobs that many women do, have multi-factorial causes and usually develop slowly, over time. Although there is now recognition of the workplace causes and successful preventive measures for both, good research and successful intervention require that all factors are examined and the real work situation is looked at. For example, stress factors that women face more frequently than men such as conflict between work and home life, sexual harassment and discrimination may be overlooked in risk assessments for work-related stress. In addition, diseases with multi-factorial causes are still less likely to be compensated financially.

Some areas of women's health in relation to work have received almost no attention, such as menstrual disorders and the menopause. In contrast, much attention has been paid to their health as expectant or new mothers, while far less attention has been given to male reproductive health and work.

Apart from reproductive capacity, there are other biological differences between women and men. Men tend to be taller, stronger, etc. Outside of the OSH field, it is recognised that women suffer health problems at lower levels of alcohol consumption than men. Men suffer testicular cancer, women breast cancer. However, as exposure due to working conditions is so strongly related to health outcomes in both women and men, studies of women and men working under the same circumstances, with the same exposures, are needed to investigate any sex differences in vulnerability.

Given the strong links between job, and therefore the level of exposure to hazards, and health outcome, individual differences become less important if the focus is on controlling risks at source, to create a safe and healthy work environment for all. The use of patient handling equipment in hospitals reduces manual handling for both women and men. We have seen that allocating jobs to women and men, on the basis of strength, for example, is problematic in various ways. 'Light' jobs are not always as light as they may seem and heavy jobs will still damage stronger people over time. The higher rates of back injury among men should be borne in mind. There are sex differences in some types of health outcomes, which need to be monitored and studied in relation to work, such as male and female cancers.

There may also be gender differences in coping with stress, for example, but there do not appear to be many gender differences in sources of stress, although the amount of exposure and its impact can vary by gender. Again, it is important to concentrate on recognition and prevention to improve the quality of work for all. For example, it is important to improve work–life balance for all, although it would particularly benefit women. It is also important to find a way to encourage a more even sharing of domestic tasks.

Flexibility is important in a number of ways to provide a safe and healthy workplace for a diverse population, matching jobs and working conditions to people, not people to jobs. This includes working-time flexibility, for example to help improve work–life balance. Flexibility in being able to take toilet and rest breaks is particularly important for pregnant workers and those going through the menopause. We have also seen that it is important to take account of individual differences in implementing solutions, such as in the provision of protective clothing and equipment, as that designed for the standard man puts some men and more women at risk.

The use of standards, for example, for equipment design and exposure to dangerous substances is an important part of the OSH-prevention programme in the EU. Here, it is important to take account of any gender differences to create standards that protect both women and men, although many standards have been set based on data from largely male populations.

As has been mentioned, to obtain an accurate understanding of the risks to women workers and for successful prevention, they and their jobs must be included in research and information collection. The questions asked, data collection methods and method of analysis are all important. Another crucial element for improving attention to recognition and prevention of work hazards to women is the involvement of women in OSH activities and decision-making at all levels. Various examples have been given in this report of the importance of involving women workers in the risk-assessment and prevention process at the workplace and looking at the real jobs and tasks done.

Gender-sensitivity and mainstreaming

Gender differences may result in unequal patterns of health risk, use of health services and different health outcomes. The findings of this report confirm that a gender-neutral approach to occupational safety and health is contributing to the maintenance of gaps in knowledge and more attention and resources being given to men's occupational health and risks in men's work, resulting in less effective prevention. The possibility of gender inequalities in the coverage and application of gender-neutral directives, OSH priority setting and resource allocation, benchmarking, standard setting and compensation arrangements have been mentioned. Occupational health policies can take account of gender differences by adopting a gender-sensitive approach, acknowledging that certain health problems are unique to, or have more serious implications for, either women or men.

Gender mainstreaming — in other words, integration of gender concerns into the analyses, formulation and monitoring of policies, programmes and projects to reduce inequalities between men and women — is central to this, and the use of gender impact assessment is an important tool in this process.

Below is a model for mainstreaming gender into the OSH system.

Action levels for mainstreaming gender into the OSH system

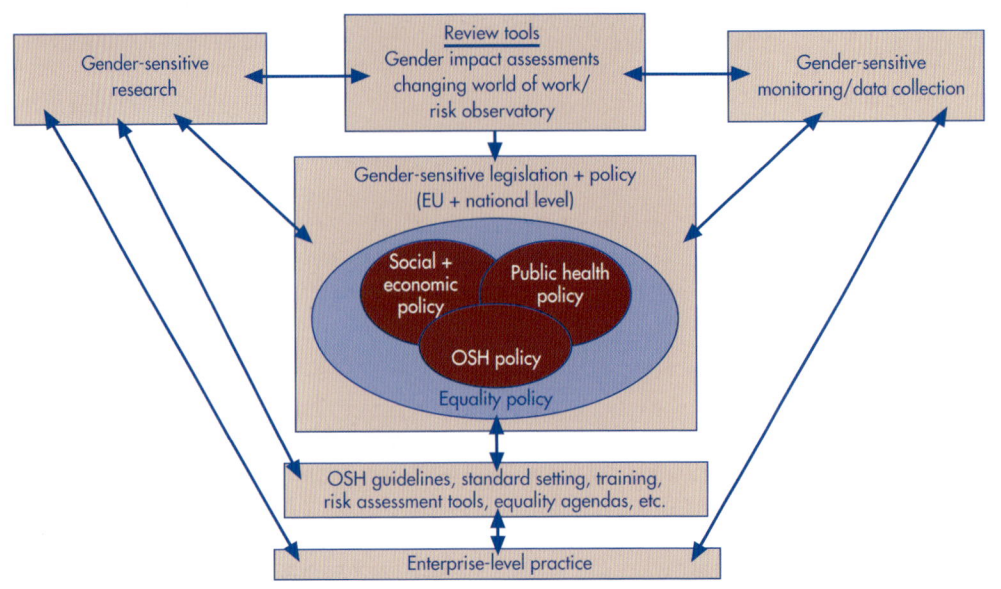

In addition, to take a holistic approach to OSH prevention requires an interdisciplinary approach between different policy fields, which is reflected in the model.

There is enough knowledge to start taking a gender-sensitive approach to OSH at all levels, from policy to practice. A gender perspective should be built into the labour inspection, development and implementation of legislation, risk assessment, and prevention practices. Occupational health professionals, both physicians and nurses as well as other experts, such as safety engineers and physiotherapists, need more training on these issues. Trade unions have a strong role to play in awareness raising, training and developing guidelines. Frameworks should be put in place to serve as a guideline for mainstreaming gender into OSH policies and practices. Experiences and practices should be shared so that approaches can be improved.

Some suggestions to take forward the agenda of mainstreaming gender into OSH are presented below.

The European Commission's Advisory Committee on Safety, Hygiene and Health could set up an ad hoc group on gender with a remit to examine practical ways to take forward the EU agenda of mainstreaming gender into OSH.

European sector dialogue committees could examine how mainstreaming and a gender-sensitive approach could be taken forward in their sector. This could include developing guidelines for risk assessment and prevention.

Another area to explore is how equalities organisations can extend their activities to include occupational safety and health. For example, the European Women's Lobby established a European women's talent bank as a resource to help increase women's participation in meetings, conferences and technical groups advising the European Commission and other European and international institutions. Initiatives such as this could be extended to include the field of occupational safety and health.

Avenues of cooperation between the Advisory Committee on Safety, Hygiene and Health and the Advisory Committee on Equalities could be explored and the Advisory Committee on Equalities could consider how they can promote equality in the OSH field within their own activities.

Similar actions can be taken at national level. In addition, OSH authorities and institutions could specifically target women's occupational health issues and areas of women's work that have been neglected, for research and intervention activities, such as has been done by NIOSH.

Organisations covering women's health, such as women's healthcare programmes or non-governmental organisations concerned with women's health issues, are an important source of information and play an important awareness-raising role. There is great scope for them to cover occupational health aspects of women's health and work with those in the OSH field, such as OSH authorities, trade unions and employers to raise awareness and develop guidance for the workplace. It has also been found that women make good use of the Internet to search for health information. Including OSH-related information on the web sites of health organisations could be a particularly useful awareness-raising method among women.

An important issue in the OSH field is to raise awareness and start prevention activities as early as possible, by taking OSH into schools and mainstreaming it into the educational process. It follows that a gender perspective should be included in this process.

Taking a gender-sensitive approach to workplace intervention

A gender-sensitive approach is needed in OSH prevention. It has been pointed out that, although the majority of European OSH legislation is gender neutral, it would be possible to apply it in a gender-sensitive way. However, guidance, awareness raising and training on taking a gender-sensitive approach would be needed to achieve this.

The key aim of a gender-sensitive risk-assessment and prevention approach is to view effectively the less visible hazards and health problems that are more common for women workers, in order to take suitable preventive action. As there are gender differences in a variety of broader issues relating to work circumstances, such as work–life conflicts, discrimination, and the involvement in decision-making in the workplace, a holistic approach to risk prevention is needed if all relevant gender differences are to be taken into account.

Risk assessment and prevention are commonly considered to cover five stages:

- Hazard identification
- Risk assessment
- Implementation of solutions
- Monitoring
- Review

All parts of the risk-assessment and prevention process need to be gender-sensitive. In Box 71 are some suggestions for making the process more gender-sensitive based on the results of this study.

Box 71: Suggestions for including gender in risk assessment

Key issues for gender-sensitive risk assessment:

✓ Having a positive commitment and taking gender issues seriously.
✓ Looking at the real working situation.
✓ Involving all workers, women and men, at all stages.
✓ Avoiding making prior assumptions about what the hazards are and who is at risk, etc.

Hazard identification, for example:

- Consider hazards prevalent in both male- and female-dominated jobs, for example covering areas highlighted in this report.
- Look for health hazards as well as safety hazards.
- Ask both women and men workers what problems they have in their work, in a structured way.
- Avoid making initial assumptions about what may be 'trivial'.
- Consider the entire workforce, for example cleaners and receptionists.
- Do not forget part-time, temporary or agency workers, and those on sick leave at the time of the assessment.
- Encourage women to report all issues that they think may affect their safety and health at work, as well as health problems that may be related to work.
- Look at and ask about the wider issues raised in this report.

Risk assessment, for example:

- Look at the real jobs being done and the real work context.
- Do not make assumptions about exposure based purely on job description or title.
- Be careful about gender bias in prioritising risks according to high, medium and low.
- Involve women workers in risk assessment. Consider using health circles and risk-mapping methods. Participative ergonomics and stress interventions can offer some approaches.
- Make sure that those doing the assessments have sufficient information and training about gender issues in OSH.

- Make sure instruments and tools used for assessment include issues relevant to both male and female workers. If they do not, adapt them.
- If outside help is used for risk assessment, inform them that they should take a gender-sensitive approach. Check that they are able to do this.
- Pay attention to any gender issues when changes are planned in the workplace and the OSH implication are looked at.

For example, for stress include:

- home–work interface, and both men's and women's work schedules;
- career development;
- harassment;
- emotional stressors;
- unplanned interruptions and doing 'several tasks at once'.

For example, for reproductive health:

- include both male and female reproductive risks;
- look at all areas of reproductive health, not just pregnant women.

For example, for musculoskeletal disorders:

- look critically at 'light work'. How much static muscle effort is involved? Does the job involve significant standing? What loads are really handled in practice, and how often?

Implementation of solutions, for example:

- Aim to eliminate risks at source, to provide a safe and healthy workplace for all workers. This includes risks to reproductive health.
- Pay attention to diverse populations and adapt work and preventive measures to workers. For example, selection of protective equipment according to individual needs, suitable for women and 'non-average' men.
- Involve women workers in the decision-making and implementation of solutions.
- Make sure women workers as well as men are provided with OSH information and training relevant to the jobs they do and their working conditions and health effects. Ensure part-time, temporary and agency workers are included.

Monitoring and review, for example:

- Make sure women workers participate in monitoring and review processes.
- Be aware of new information about gender-related occupational health issues.

Health surveillance can be part of both risk assessment and monitoring:

- Include surveillance relevant to jobs of both male and female workers.
- Take care when making assumptions, for example based on job title, about who to include in monitoring activities.

Accident records are an important part of both risk assessment and monitoring:

- Encourage the recording of occupational health issues as well as accidents.

General

- Review safety policy. Specifically, include commitment to gender mainstreaming, and relevant objectives and procedures.
- Seek to ensure that both internal and external occupational health services used will take a gender-sensitive approach.
- Provide relevant training and information on gender issues regarding safety and health risks to risk assessors, managers and supervisors, trade union representatives, safety committees, etc.
- Link occupational safety and health into any workplace equality actions, including equality plans.
- Look at ways to encourage more women to be involved in safety committees. For example, are meetings held at times when women can attend?

European Agency for Safety and Health at Work

RESEARCH

pational health physicians and nurses and others, such as safety engineers and physiotherapists, need more training on these issues. Awareness of the need for a gender-sensitive approach and support in how to achieve it are needed at all levels. Existing approaches and experiences need to be shared, and a comprehensive framework that serves as a guide for mainstreaming gender into OSH policies and practices needs to be developed. This report and the models presented in it could be used as a starting point for further discussion and activities.

3. **Taking account of all risks.** Some risks faced by women have been given less attention, such as standing work and reproductive health issues other than pregnancy. More research is needed in the neglected areas and they need to be included in assessment and prevention activities. A holistic approach to OSH should be taken that recognises and takes account of issues such as conflict between work and home life, harassment and discrimination.

4. **Taking account of real work situations.** Gender differences in exposures to hazards, arising from gender segregation in employment, have a strong impact on gender differences in occupational health outcomes. The work-based causes and requirements for prevention even for upper limb disorders and stress are well documented. However, it is very important that the real tasks being done, and the real work context are examined both for successful prevention and for research and monitoring.

8.

RECOMMENDATIONS

Key issues

1. **Prioritising prevention.** Continuous efforts are needed to improve working conditions and improve work organisation so that they are suited to both women and men.

2. **Promoting and facilitating a gender-sensitive approach**. Gender-sensitivity is needed in research, policy and prevention practice to help ensure effective prevention and avoid gender bias in occupational safety and health (OSH). A gender perspective should be systematically built into labour inspection, legislation and guidance, and risk-assessment and prevention practices. Occu-

General

1. **Promoting work–life balance policies.** Women generally carry out a larger share of household duties and family caring responsibilities. Many older women care for elderly relatives. Improved social support for carers

139

would particularly benefit women, but a more equal distribution of home responsibilities can be positive for both men and women. Men are more likely to work long full-time hours, and in this case any non-standard or irregular working schedules have a greater impact on incompatibility between work and life. Many of those working long hours would like to shorten their hours, and many of those working shorter hours would like the opportunity to increase the number of hours worked. Work–life balance policies, including family-friendly elements, suitable hours and flexibility, should be initiated in European workplaces for the benefit of both women and men, and they should take account of both women's and men's work schedules and be designed to be attractive to both. More sharing of existing good practice and more intervention research are needed in the workplace on work–life balance and family-friendly strategies, flexitime, job sharing, etc.

2. **Paying attention to specific groups of women workers.** Women workers are not a homogeneous group, nor are, for example, older workers or male workers. More attention should be paid to the working situation and occupational safety and health needs of women in non-traditional jobs, young and older women workers, commercial sex workers and other vulnerable groups.

3. **Promoting interdisciplinary cooperation.** The multi-factorial nature of gender differences in occupational safety and health require active cooperation among different areas of policies, research, education and information. Ways of promoting this cooperation need to be investigated and good practices shared.

4. **Mainstreaming occupational safety and health into equality agendas at all levels.** For example, by including occupational safety and health in all health equality agendas, or by incorporating occupational safety and health into workplace equality plans.

5. **Reducing discrimination and improving the quality of women's employment.** These important issues are largely outside the scope of this report. For strategies for improvement in this area, such as provisions directed at enhancing women's job access, reducing sex segregation and diminishing pay discrimination see, for example, Annex 14, taken from European Foundation (2002c).

6. **Improving women's participation in occupational safety and health decision-making.** Women should be represented and directly involved in the decision-making concerning occupational safety and health. Women's own views, their experiences, knowledge and skills should be reflected when formulating and implementing health-promotion strategies within occupational safety and health. Ways of achieving this should be explored and existing good practices exchanged.

Research and OSH monitoring

1. **Basing exposure assessment on the real jobs people do.** Exposure assessment should be improved and not based on job title, but on the real work situation and tasks people carry out. Possible synergistic and cumulative effects of chemicals among themselves and with other hazards should be included.

2. **Assessing epidemiological research for bias.** Existing epidemiological research should be critically assessed to find any systematic bias in the way investigations are done when studying women's occupational safety and health.

3. **Routinely and systematically including the gender dimension in data collection and statistical monitoring.** This includes the collection of national reportable occupational accident, disease and ill-health data, as well as record keeping in relation to compensation data. Gender analyses of the information collected by Eurostat and through the European surveys on working conditions, for example, are very useful. However, it is important to consider whether the information collected and the indicators available are adequate for identifying the particular features that typically characterise women's employment. The recommendations from the gender analyses of the third European survey on working conditions for modifications to the survey should be implemented. The further development of Member State occupational safety and health statistics by gender is needed to provide more information in general and on any differences among the Member States. The quality of gender statistics on occupational safety and health outcomes can be enhanced if they take account of the number of hours worked and the division into occupations and sectors. The closer the breakdown according to the real tasks people carry out, the better. Practical means of improving the quality of occupational safety and health statistics by gender should be investigated.

4. **Investigating and comparing Member State differences further.** Further investigation is needed of gender differences in working conditions and occupational outcomes within individual Member States, in order to see if and why there are any differences among Member States, especially in those where data are lacking.

5. **Gathering information from the candidate countries.** There are differences regarding gender and working conditions between candidate countries and the Member States and among the different candidate countries. It would be useful to look in more detail at policies and practices from the candidate countries.

6. **Redressing the gender imbalance in research programmes.** Less research attention has been given to occupational risks to women and to jobs dominated by women. Where necessary, OSH research programmes should specifically include topics relevant to women and their workplaces. More intervention studies on prevention of occupational risks to women and mainstreaming gender into prevention activities are needed.

7. **Considering the impact of changes in the world of work on gender and occupational safety and health.** Changes in the world of work have implications for occupational safety and health. A recent example has been the introduction of teleworking. Changes should be assessed for their occupational safety and health implications, and this should include a gender assessment.

Policy development

1. **Mainstreaming gender into occupational safety and health strategy.** Gender should be mainstreamed into all areas of occupational safety and health at EU and national level, as outlined in the European Community strategy 2002–06. The gender element should be well defined and transparent and a comprehensive framework should be created that will serve as a guide for mainstreaming gender into occupational safety and health policies, programmes and practices. The Commission and the Advisory Committee on Safety, Hygiene and Health (ACSHH) could support this element of the strategy, for example, by specifically

addressing the gender issue in current activities on musculoskeletal disorders, workplace violence and hospitals, as these are risks and work areas of significance to many women workers, as well as in activities on occupational diseases. They could also consider whether there is a need to set up a specific ACSHH group on mainstreaming gender into OSH. Member States should also look at what practical steps they can take to mainstream gender into their occupational safety and health activities and there should be exchange of existing experiences and practices in this area.

2. **Applying gender impact assessments as a mainstreaming tool.** In accordance with the EU strategy on promoting equality, gender impact assessments should be carried out when existing EU occupational safety and health directives, guidelines, policies and actions are reviewed or new ones are being developed. This should include prevention, standardisation and compensation, as well as information collection and access to OSH services and resources.

3. **Incorporating gender and occupational working conditions into standard setting.** Women's health and safety may be less well protected than men's because many occupational safety and health standards and exposure limits to hazardous substances have been based on male populations and laboratory tests. Data on both sexes and gender differences in working conditions should be taken into account in European and international standardisation.

4. **Mainstreaming OSH gender issues into other policy areas.** OSH gender issues should be mainstreamed into other policy areas such as health policy or actions on corporate social responsibility (CSR), for example.

Awareness raising and good practice in prevention

1. **Improving gender-sensitivity in the implementation of existing directives etc.** The rather broad requirements of risk assessment and prevention in the gender-neutral European Union occupational safety and health directives and guidelines do not exclude taking a more gender-sensitive approach in their implementation. However, more gender-sensitive methods for implementation need to be developed. This could include incorporating information into guidelines or codes of practice that the Commission and Member States produce to support the implementation of the legislation as well as producing specific guidance, or training and information for labour inspectorates. Social partners have an important role to play, for example to develop guidelines. All the various occupational safety and health actors need information and training about women's working conditions.

2. **Investigating and sharing good practices.** This report provides some examples of approaches and practices in individual Member States and from other sources. It would be useful to gather and share more examples of policies, programmes and good practices from the workplace from all Member State countries.

3. **Awareness raising and promoting good practices and methods for including gender in risk assessment and involving women workers.** Active promotion of a gender-sensitive approach to risk assessment in the workplace is needed. This would include awareness raising and training, backed by appropriate information for the workplace. The Commission and the Advisory Committee on Safety, Hygiene and Health, Member States and social partners should explore how they can support this recommendation.

9. GLOSSARY

Sex and gender

The existing differences between men and women are of a biological and social nature:
- **Sex** refers to the biologically determined differences between men and women, which are universal.
- **Gender** refers to the social differences between women and men that are learned, changeable over time and have wide variations both within and between cultures.

Example: While only women can give birth (biologically determined), biology does not determine who will raise the children (gendered behaviour).

Source of definition: European Commission (no date), 'A guide to gender impact assessment'.

Gender equality — equality between women and men

1. Gender equality means that all human beings are free to develop their personal abilities and make choices without the limitations set by strict gender roles; that the different behaviour, aspirations and needs of women and men are equally valued and favoured. Formal (*de jure*) equality is only a first step towards material (de facto) equality. Unequal treatment and incentive measures (positive action) may be necessary to compensate for past and present discrimination. Gender differences may be influenced by other structural differences, such as race/ethnicity and class. These dimensions (and others, such as age, disability, marital status and sexual orientation) may also be relevant.

    Source of definition: European Commission (no date), 'A guide to gender impact assessment'.

2. Absence of discrimination on the basis of a person's sex in opportunities and the allocation of resources or benefits or access to services.

    Source of definition: WHO (1998), Gender and health technical paper.

Gender blindness or gender neutrality

Failure to recognise that gender is an essential determinant of social outcomes.

Source of definition: WHO (1998), Gender and health technical paper.

Gender-sensitivity

Ability to perceive existing gender differences, issues and inequalities and incorporate these into strategies and actions.

Source of definition: WHO (1998), Gender and health technical paper.

Gender impact assessment — gender analysis

1. Gender impact assessment means to compare and assess, according to gender-relevant criteria, the current situation and trend with the expected development resulting from the introduction of the proposed policy.

   Source of definition: European Commission (no date), 'A guide to gender impact assessment'

2. Gender analysis examines the differences and disparities in the roles that women and men play, the power imbalances in their relations, their needs, constraints and opportunities and the impact of these differences on their lives. In health, a gender analysis examines how these differences determine differential exposure to risk, access to the benefits of technology, information, resources and healthcare and the realisation of rights. A gender analysis must be done at all stages of an intervention, from priority setting and data collection to the design, implementation and evaluation of policies or programmes.

   Source of definition: WHO (1998), Gender and health technical paper.

Mainstreaming — gender equality mainstreaming

1. In the European Commission communication on mainstreaming (COM (96) 67), mainstreaming is defined as 'not restricting efforts to promote equality to the implementation of specific measures, but mobilising all general policies and measures specifically for the purpose of achieving equality'. The gender and equality dimension should be taken into account in all policies and activities, in the planning, implementation, monitoring and evaluation phases.

   Source of definition: European Commission (no date), 'A guide to gender impact assessment'.

2. The integration of gender concerns into the analyses, formation and monitoring of policies, programmes and projects, with the objective of ensuring that these reduce inequalities between women and men.

   Source of definition: WHO (1998), Gender and health technical paper.

Mainstreaming — occupational safety and health mainstreaming

Mainstreaming aims to make risk management principles and 'occupational safety and health thinking' an intrinsic part of the way decisions are made and actions are taken in the workplace, so that health and safety is not just an 'add-on'.

Source of definition: European Agency, 'Forum 8 — Learning about occupational safety and health'.

Equality plans

Plans designed by management to incorporate an equality perspective and programme into all company policies.

Source of definition: European Foundation (2002c).

10. REFERENCES

ACTU (2003), *National survey of workplace issues 2002*, Australian Council of Trade Unions, Melbourne (http://www.actu.asn.au/public/news/1043965764_16041.html).

Ahjo nro (2002), 'Uudenkaupungin autotehtaalle tasa-arvopalkinto' ('Uusikaupunki car factory wins equality award'), *Ahjo nro* 21/2002, Helsinki (http://www.metalliliitto.fi/21–02tasapalk.htm).

Akyeampong, E. B. (1992), 'L'absentéisme', *Tendances Sociales Canadiennes* 25, pp. 25–28.

Albarracan, D. (2002), *El sector de servicio doméstico: mujer e inmigración*, CIREM Foundation, European Industrial Relations Observatory web site (2002), *Domestic work examined* (http://www.eiro.eurofound.ie/2002/05/Feature/ES0205206F.html), European Foundation for the Improvement of Living and Working Conditions.

Alterman, T., Shekelle, R. B., Vernon, S. W. and Burau, K. D. (1994), 'Decision latitude, psychological demand, job strain and coronary heart disease in the Western Electric Study', *American Journal of Epidemiology* 139(6), pp. 620–627.

American Psychological Association (1996), 'Research agenda for psychosocial and behavioural factors in women's health', American Psychological Association, Recommendations.

Ammattitaudit 2000, 'Työperäisten sairauksien rekisteriin ilmoitetut uudet tapaukset' (Occupational diseases 2000. 'Reported new cases of occupational diseases'), Työterveyslaitos, Helsinki, 2001.

Ammattitaudit 2001, 'Työperäisten sairauksien rekisteriin ilmoitetut uudet tapaukset'. (Occupational diseases 2001. 'Reported new cases of occupational diseases').Työterveyslaitos, Helsinki, 2002.

Andersen, A., Barlow, L., Engeland, A., Kjaerheim, K., Lynge, E. and Pukkala, E. (1999), 'Work-related cancer in the Nordic countries', *Scandinavian Journal of Work and Environmental Health*, 25, Supplement 2, pp. 1–116.

Anderson, B. and Phizacklea, A. (1997), *Migrant domestic workers. A European perspective*, Report for the Equal Opportunities Unit, DG V, European Commission.

AOK Bayern (2002), *Betriebliche Gesundheitsförderung in Klein- und Mittelbetrieben — Ein Modellprojekt der AOK Bayern*.

Ariëns, G. A. M., Bongers, P. M., Douwes, M., Miedema, M. C., Hoogendoorn, M. E., Wal, G., van der Bouter, L. M. and van Mechelen, W. (2001), 'Are neck flexion, neck rotation and sitting at work risk factors for neck pain? Results of a prospective cohort study in an occupational setting', *Occupational and Environmental Medicine* 58, pp. 200–207.

Aronsson, G. and Gustafsson, K. (2002), 'Semester — Forfarande en arbetarskyddsfråga? En empirisk studie av semester och återhämtning' ('Vacation — Still a question of occupational safety? An empirical study of vacation and return to work'), *Arbetsmarknad and Arbetsliv*, årg 8, nr 2, summer, 2002.

Artazcoz, L./Institut Municipal de Salut Publica (2001), 'Housework, paid work and women's health', Presented at 'A seminar on the gender dimension in health and safety', 16 November, Belgium EU Presidency/Federal Ministry of Labour and Employment Equal Opportunities Unit with the TUTB, Brussels.

Atkinson, J. (2000), *Employment options and labour market participation*, European Foundation for the Improvement of Living and Working Conditions, Office for Official Publications of the European Communities, Luxembourg.

Autio, L., Heloma, A., Taskinen, K. and Reijula, K. (2001), 'Ravintolatyöntekijöiden tupakointi ja tupakansavu-altistus — Toteutuuko tupakkalaki?' ('Restaurant employees' smoking and their exposure to cigarette smoke — Does the anti-smoking law go ahead?'), *LSS* 2001 56, pp. 5289–5292.

Auvinen, A., Hietanen, M., Luukkonen, R. and Koskela, R.-S. (2002), 'Brain tumours and salivary gland cancers among cellular telephone users', *Epidemiology* 13(3), pp. 356–359.

Axelsson, A. and Prasher, D. (2000), 'Tinnitus induced by occupational and leisure noise', *Noise and Health* 8, pp. 47–54.

Axelsson, G., Ahlborg, G. and Bodin, L. (1996), 'Shift work, nitrous oxide exposure and spontaneous abortion among Swedish midwives', *Occupational and Environmental Medicine* 53, pp. 374–378.

Ayuso-Mateos, J. L., Vazquez-Barquero, J. L. and Dowrick, C. (2001), 'Depressive disorders in Europe: prevalence figures from the ODIN (European Outcome of Depression International Network) study 2001', *British Journal of Psychiatry* 179, pp. 308–316.

Baird, D. D. and Strassmann, B. I. (2000), 'Women's fecundability and factors affecting it', in Goldman, M. and Hatch, M. (eds), *Women and Health*, Academic Press, San Diego, San Francisco, New York, Boston, London, Sydney, Tokyo.

Bakker, A., Schaufeli, W. B. and van Dierendonck, D. (2000), 'Burnout: prevalence, risk groups and risk factors', in Houtman, I., Schaufeli, W. B. and Taris, T. (eds), *Mental fatigue and work* (in Dutch), Samsom, Alphen a/d Rijn, pp. 65–82.

Ballard, T., Corradi, L., Lauria, L., Mazzanti, C., Sgorbissa, F. and Romito, P. (2002a), 'Flight attendants talk about work, family and health: generating hypotheses from qualitative research', Workshop TuW15, in Bildt, C., Gonäs, L., Karlqvist, L. and Westberg, H. (eds), *Women, work and health — IIIrd international congress — Book of abstracts*, Arbetslivsinstitutet, Stockholm.

Ballard, T., Lagorio, S., De Santis, M., De Angelis, G., Santaquilani, M., Caldora, M. and Verdecchia, A. (2002b), 'A retrospective cohort mortality study of Italian commercial airline cockpit crew and cabin attendants 1965–96', *International Journal of Occupational Environmental Health* 8(2), pp. 87–96.

Barnett, R. C. and Goreis, K. C. (2000), 'Reduced hours, job role quality and life satisfaction among married women physicians with children', *Psychology and Women Quarterly* 24, pp. 358–364.

Barrenäs, M. (1998), 'Pigmentation and noise-induced hearing loss: Is the relationship between pigmentation and noise-induced hearing loss due to an otoxic pheolaminin interaction or to otoprotective eumelan effects. Advances in noise research', in Prasher, D. and Luxon, L. (eds.), *Biological effects of noise*, Volume 1, Whurr Publishers Ltd, London, pp. 59–70.

Barros Duarte, C., Ramos, S. and Lacomblez, M. (2002), 'From work to occupational health: working conditions' analysis in the textile and clothing sector — the particularity of women's work', Workshop TuW09:2, in Bildt, C., Gonäs, L., Karlqvist, L. and Westberg, H. (eds), *Women, work and health — IIIrd international congress — Book of abstracts*, Arbetslivsinstitutet, Stockholm.

Baruch, G. K., Biener, L. and Barnett, R. C. (1987), 'Women and gender in research on work and family stress', *American Psychologist* 42, pp. 130–136.

Bastian, R. W. and Thomas, M. D. (2000), *Talkativeness and vocal loudness: Do they correlate with laryngeal pathology? — A study of the vocal overdoer/underdoer continuum*, Voice Foundation meeting, Philadelphia, PA 6/30/00, Voicedoctor.net (http://www.voicedoctor.net/ physician/ article/talk/abstract.html).

Batra, P. (2002), 'Aspects of gender relations in Greek industry', Workshop TuW09:4, in Bildt, C., Gonäs, L., Karlqvist, L. and Westberg, H. (eds), *Women, work and health — IIIrd international congress — Book of abstracts*, Arbetslivsinstitutet, Stockholm.

Benach, J., Gimeno, D. and Benavides, F. G. (2002), *Types of employment and health in the European Union*, European Foundation for the Improvement of Living and Working Conditions, Office for Official Publications of the European Communities, Luxembourg.

Benschop, Y. and Doorewaard, H. (1996), 'Lood om oud ijzer. De gendersubtekst van tayloristische en sociotechnische organisaties' ('Six of one and half a dozen of the other. The gender subtext of Tayloristic and sociotechnical organisations'), *Tijdschrift voor Arbeidsvraagstukken* 12(3), pp. 238–250.

Bercusson, B. and Dickens, L. (1996), *Equal opportunities and collective bargaining in Europe. Part 1: Defining the issues*, European Foundation for the Improvement of Living and Working Conditions, Office for Official Publications of the European Communities, Luxembourg.

Bercusson, B. and Weiler, A. (1999), *Equal opportunities and collective bargaining in Europe. Part 3: Analysis of agreements*, European Foundation for the Improvement of Living and Working Conditions, Office for Official Publications of the European Communities, Luxembourg.

Berger, E., Royster, L. and Thomas, W. (1978), 'Presumed noise-induced permanent threshold shift resulting from exposure to an A-weighted Leq of 89 dB', *Journal of the Acoustical Society of America* 64(1), pp. 192–197.

Bergman, B. P. and Miller, S. A. (2001), 'Equal opportunities, equal risks? Overuse injuries in female military recruits', *Journal of Public Health Medicine* 23(1), pp. 35–39.

Bielenski, H. and Hartmann, J. (2000), *Combining family and work: the working arrangements of women and men*, European Foundation for the Improvement of Living and Working Conditions, Office for Official Publications of the European Communities, Luxembourg.

Bielenski, H., Bosch, G. and Wagner, A. (2002), *Working time preferences in 16 European countries*, European Foundation for the Improvement of Living and Working Conditions, Office for Official Publications of the European Communities, Luxembourg.

BIFU — Banking, Insurance and Finance Union (now UniFI), *Occupational voice loss. A negotiator's guide*, London.

Bildt, C. and Michélsen, H. (2002), 'Gender differences in the effects from working conditions on mental health: a four-year follow-up', *International Archives of Occupational and Environmental Health* 75, pp. 252–258.

Bildt, C., Gonäs, L., Karlqvist, L. and Westberg, H. (eds) (2002), *Women, work and health — IIIrd international congress — Book of abstracts*, Arbetslivsinstitutet, Stockholm (http://www.arbetslivsinstitutet.se/wwh/default.asp).

Bildt Thorbjörnsson, C. and Lindelöw, M. (1998), 'Psychiatric ill health and conditions at work', in Kilbom, Å., Messing, K. and Bildt Thorbjörnsson, C. (eds), *Women's health at work*, Arbetslivsinstitutet, Stockholm.

Black, D. W., Doebbeling, B. N., Voelker, M. D., Clarke, W. R., Woolson, R. F., Barrett, D. H. and Schwartz, D. A. (2000), 'Multiple chemical sensitivity syndrome: symptom prevalence and risk factors in a military population', *Archives of International Medicine* 160(8), pp. 1169–1176.

Blair, A. (1998), 'Occupational cancer among women, an overview', *Women's health, occupation, cancer and reproduction*, 14 to 16 May, Reykjavik, Iceland.

Blatter, B. M. and Bongers, P. M. (2002), 'Duration of computer use and mouse use in relation to musculoskeletal disorders of neck or upper limb', *International Journal of Industrial Ergonomics* 30, pp. 295–306.

Bleijenbergh, I., de Bruijn, J. and Dickens, L. (1999), *Equal opportunities and collective bargaining in Europe. Part 5: Strengthening and mainstreaming equal opportunities through collective bargaining*, European Foundation for the Improvement of Living and Working Conditions, Office for Official Publications of the European Communities, Luxembourg.

Blikvaer, T. and Helliesen, A. (1997), *Sickness absence. A study of 11 LES countries*, Luxembourg income study, LES Working Paper 3, Luxembourg (http://www.lisproject.org/publications/leswps/leswp3.pdf).

Boice, J. D. and McLaughlin, J. (2002), *Epidemiological studies of cellular telephones and cancer risk*, Statens srålskyddsinstitut, SSI Rapport 2002, p. 16.

Bongers, P. M., de Winter, C. R., Kompier, M. A. J. and Hildebrandt, V. H. (1993), 'Psychosocial factors at work and musculoskeletal disease', *Scandinavian Journal of Work, Environment and Health* 19, pp. 297–312.

Borg, A., Canlon, B. and Engström, B. (1992), 'Individual variability of noise-induced hearing loss', in Dancer, A. L., Henderson, D., Salvi, R. J. and Hamernik, R. P. (eds), *Noise-induced hearing loss*, Mosby Year Book, St Louis, pp. 467–475.

Borras, J., Borras, J. M., Galceran, J., Sanchez, V., Moreno, V. and Gonzalez, J. R. (2001), 'Trends in smoking-related cancer incidence in Tarragona, Spain, 1980–96', *Cancer Causes Control* 12(10), pp. 903–908.

Borrill, C. and Kidd, J. M. (1994), 'New parents at work: jobs, families and the psychological contract', *British Journal of Guidance and Counselling* 22(2), pp. 219–231.

Bosma, H., Marmot, M. G., Hemingway, H., Nicholson, A. G., Brunner, E. and Stansfeld, A. (1997), 'Low job control and risk of coronary heart disease in Whitehall II (prospective cohort) study', *British Medical Journal* 314, pp. 558–565.

Boyd, C. (2002), 'Customer violence and employee health and safety', *Work, Employment and Society* 16(01), pp. 151–170.

Brisson, C., Laflamme, N., Moisan, J., Milot, A., Msse, B. and Vzina, M. (1999), 'Effect of family responsibilities and job strain on ambulatory blood pressure among white-collar women', *Journal of the American Psychosomatic Society* 61(2), p. 205.

Broberg, E. (2001), 'Serious occupational accidents', in Marklund, S. (ed.), *Worklife and health in Sweden 2000*, Arbetslivsinstitutet, Stockholm.

Buckle, P. and Devereux, J. (1999), *Work-related neck and upper limb musculoskeletal disorders*, European Agency for Safety and Health at Work, Office for Official Publications of the European Communities, Luxembourg.

Budig, J. M. and England, P. (2001), 'The wage penalty for motherhood', *American Sociological Review* 66, pp. 204–225.

Bullinger, M., Morfeld, M., von Mackensen, S. and Brasche, S. (1999), 'The sick-building syndrome — Do women suffer more?', *Zentralblatt für Hygiene und Umweltmedizin,* 1999, Aug. 202 (2–4), pp. 235–241.

Bylund, S. H. and Burström, L. (2003), 'Power absorption in women and men exposed to hand-arm vibration', *International Archives of Occupational and Environmental Health* 77.

Bylund, S. H., Burström, L. and Knutsson, A. (2002), 'A descriptive study of women injured by hand-arm vibration', *Annals of Occupational Hygiene* 46(3), pp. 299–307.

Campo, P. and Lataye, P. R. (1992), 'Intermittent noise and equal energy hypotheses', in Dancer, A. L., Henderson, D., Salvi, R. J. and Hamernik, R. P. (eds), *Noise-induced hearing loss*, Mosby Year Book, St Louis, pp. 456–466.

Cancedda, A. (2001), *Employment in household services*, European Foundation for the Improvement of Living and Working Conditions, Office for Official Publications of the European Communities, Luxembourg.

Castells, M. (1996), 'The rise of the network society. The information age', *Economy, society and culture*, Volume 1, Blackwell, Massachusetts, Oxford.

Cinbiose, 'Improving the health of women in the workforce — Action plan developed by the meeting of Canadian researchers and representatives of women workers, 26 to 28 March 1998, at the Université du Québec à Montréal', University of Quebec (http://www.unites.uqam.ca/cinbiose/anglais/pub/pub.actionplan.html).

Clarke, S. (2001), 'Earnings of men and women in the EU: the gap narrowing but only slow', Theme 3: Population and social conditions, Eurostat (http://europa.eu.int/comm/employment_social/equ_opp/statistics_en.html).

Cockburn, C. (1983), *Brothers. Male dominance and technological change*, Pluto Press, London.

Cockburn, C. (1986), *Machinery of dominance*, Pluto Press, London.

Comisiones Obreras (CC.OO.) (2001), *Salud-Laboral: diferencia género*, Federación de Comunicación y Transporte de CC.OO., Madrid.

Cooper, C. L. (1998), 'The changing psychological contract at work', *Work and Stress* 12(2), pp. 97–100.

Cooper, C. L. and Swanson, N. (2002), *Workplace violence in the health sector — State of the art*, ILO, WHO, ICN and PSI, International Labour Office, Geneva.

Cooper, C. L., Hoel, H. and di Martino, V. (2003), *Preventing violence and harassment in the workplace*, European Foundation for the

Improvement of Living and Working Conditions, Office for Official Publications of the European Communities, Luxembourg.

Cooper, G. S., Savitz, D. A., Millikan, R. and Chiu Kit, T. (2002), 'Organochlorine exposure and age at natural menopause', *Epidemiology* 13(6), pp. 729–733.

Cortazar Lopez, M., Kareaga Uriate, G., Lansac Aquilué, M., Irusta Onandia, J. A. and Azuara Blanco, S. (2002), *Estudio de los trastornos de voz en docents y auxiliaries de educación especial de al enseñanza pública de Bizkaia*, Servicio Médico de la Delegación de Educación de Bizkaia del Gobierno Vasco, Bilbao.

Council Directive 76/207/EEC of 9 February 1976 on the implementation of the principle of equal treatment for men and women as regards access to employment, vocational training and promotion and working conditions, OJ L 39, 14.2.1976, p. 40, Derogation in 194N, Incorporated by OJ L 1, 3.1.1994, p. 484, Luxembourg.

Council Directive 89/391/EEC of 12 June 1989 on the introduction of measures to encourage improvements in the safety and health of workers at work, OJ L 183, 29.6.1989, Luxembourg, pp. 1–8.

Council Directive 89/654/EEC concerning the minimum safety and health requirements for the workplace, OJ L 393, 30.12.1989, Luxembourg, pp. 1–12.

Council Directive 89/655/EEC concerning the minimum safety requirements for the use of work equipment by workers at work, OJ L 393, 30.12.1989, Luxembourg, pp. 13–17.

Council Directive 89/656/EEC on the minimum safety requirements for the use of personal protective equipment, OJ L 393, 30.12.1989, Luxembourg, pp. 18–28.

Council Directive 92/85/EEC on the introduction of measures to encourage improvements in the safety and health at work of pregnant workers and workers who have recently given birth or are breastfeeding, OJ L 348, 28.11.1992, Luxembourg, pp. 1–8.

Council Directive 2002/73/EC of 23 September 2002 amending Council Directive 76/207/EEC on the implementation of the principle of equal treatment for men and women as regards access to employment, vocational training and promotion and working conditions, OJ L 269, 5.10.2002, Luxembourg, pp. 15–20.

Cousins, C. (2000), 'Women and employment in southern Europe: the implications of recent policy and labour market directions', *South European Society and Politics* 5(1) (summer 2000), pp. 97–122.

Cox, T. and Rial-González, E. (2002), 'Work-related stress: the European picture', *Working on stress*, Magazine 5 of the European Agency for Safety and Health at Work, Office for Official Publications of the European Communities, Luxembourg.

Cox, T., Griffiths, A. and Rial-González, E. (2000), *Research on work-related stress*, European Agency for Safety and Health at Work, Office for Official Publications of the European Communities, Luxembourg.

Coyle, A. (1982), 'Sex and skill in the organisation of the clothing industry', in West, J. (ed.), *Work, women and the labour market*, Routledge and Kegan Paul, London, pp.10–27.

Cuelenaere, B. (1997), *Going to work after long-term sick leave. A study about rehabilitation in men and women* (in Dutch), Erasmus University, Rotterdam.

Current Intelligence Bulletin 57 (1996), 'Violence in the workplace. Risk factors and prevention strategies', National Institute for Occupational Safety and Health, Cincinnati.

Danish Agency for Trade and Industry, 'Women entrepreneurs now and in the future' (http://www.efs.dk/publikationer/rapporter/womenentr/kap01.html).

Dassen, T. W. N., Nijhuis, F. J. N. and Philipsen, H. (1990), 'Career prospects for male intensive care nurses' (in Dutch), *Gedrag Organisatie* 1, pp. 32–47.

Daubas-Letourneux, V. and Thébaud-Mony, A. (2002), *Organisation du travail et santé dans l'Union européenne*, European Foundation for the Improvement of Living and Working Conditions, Office for Official Publications of the European Communities, Luxembourg (http://www.eurofound.eu.int/publications/files/EF0206FR.pdf).

Davidson, M. J., Cooper, C. L. and Baldini, V. (1995), 'Occupational stress in female and male graduate managers — A comparative study', *Stress Medicine* 11, pp. 157–175.

Davis, S., Mirick, D. K. and Stevens, R. G. (2001), 'Night shift work, light at night and risk of breast cancer', *Journal of the National Cancer Institute* 93(20), pp. 1557–1562.

Davis, A., Smith, P. and Wade, A. (1998), *A longitudinal study of hearing — Effects of age, sex and noise*, Proceedings of Nordic Noise, 12 to 15 March 1998, Stockholm.

de Lange, A., Kompier, M. A. J., de Jonge, J., Taris, T. and Houtman, I. (2001), 'Hoogwaardige longitudinaal vragenlijstonderzoek en het demand-control-support model' ('High-quality longitudinal survey research and the demand-control-support model'), *Gedrag Organisatie* 14(5), pp. 254–272.

Devereux, J. (2000), 'Work-related stress and MSDs: Is there a link?', *Preventing work-related musculoskeletal disorders*, Magazine 3 of the European Agency for Safety and Health at Work, Office for Official Publications of the European Communities, Luxembourg, p. 19, (http://agency.osha.eu.int/publications/magazine/3/en/mag3_en.pdf).

Dhondt, S., Goudswaard, A. and Knave, B. (2002), *Research on changing world of work*, European Agency for Safety and Health at Work, Office for Official Publications of the European Communities, Luxembourg.

Diamantopoulou, A. (2002), 'Europe under stress', *Working on stress*, Magazine 5 of the European Agency for Safety and Health at Work, Office for Official Publications of the European Communities, Luxembourg, p. 3.

Dickens, L. (1998), *Equal opportunities and collective bargaining in Europe. Part 4: Illuminating the process*, European Foundation for the Improvement of Living and Working Conditions, Office for Official Publications of the European Communities, Luxembourg.

Dikkers, F. G. and Nikkels, P. G. (1995), 'Benign lesions of the vocal folds: histopathology and phonotrauma', *Annals of Otology, Rhinology and Laryngology* 104 (9 Pt 1), pp. 698–703.

di Martino, V. (2002), *Workplace violence in the health sector — Country case studies (Brazil, Bulgaria, Lebanon, Portugal, South Africa, Thailand and an additional Australian study)*, Synthesis report, ILO, WHO, ICN and PSI, International Labour Office, Geneva.

di Martino, V., Hoel, H. and Cooper, C. L. (2002), *Preventing violence and harassment in the workplace*, European Foundation for the Improvement of Living and Working Conditions, Office for Official Publications of the European Communities, Luxembourg.

Douillet, P. and Aptel, M. (2001), 'Preventing MSDs: towards a global approach', *Preventing* work-related musculoskeletal disorders, Magazine 3 of the European Agency for Safety and Health at Work, Office for Official Publications of the European Communities, Luxembourg, pp. 4–6.

Doyal, L. (2002), *The health and work of older women — A neglected issue,* Pennell Trust, Huddersfield (www.pennellwomenshealth.org).

Dreyer, L., Andersen, A. and Pukkala, E. (1997), 'Occupation', *Acta Microbiologica et Immunologica Scandinavica* (APMIS), 1997, Supplement No 76, Vol. 105, 'Avoidable cancers in the Nordic countries', pp. 68–79.

Dulk, L. den (2002), 'Employers and the caring employee: how Dutch, Italian, British and Swedish organisations facilitate the combination of work and care' (in Dutch), *Gedrag Organisatie* 15(4), pp. 225–239.

Dupré, D. (2001), 'Work-related health problems in the EU 1998–99', *Statistics in Focus*, Theme 3: Population and social conditions, 17/2001, Eurostat, Office for Official Publications of the European Communities, Luxembourg.

Dupré, D. (2002), 'The health and safety of men and women at work', *Statistics in Focus*, Theme 3: Population and social conditions, 4/2002, Eurostat, Office for Official Publications of the European Communities, Luxembourg.

Eklund, I., Englund, A. and Wikman, A. (2001), 'Working conditions in Sweden and Europe', in Marklund, S. (ed.), *Worklife and health in Sweden 2000*, Arbetslivsinstitutet, Stockholm.

Ekstedt, E. (1999), 'Forms of employment in a project-intensive economy', *American Journal of Industrial Medicine,* Supplement 1.

Ekstedt, E., Lundin, R. A., Soderholm, A. and Wirdenius, H. (1999), *Neo-industrial organising. Renewal by action and knowledge formation in a project-intensive economy*, Routledge, London and New York.

End abuse, family violence prevention fund (http://endabuse.org/programs/workplace/).

Endometriosis.org web site (http://www.endometriosis.org/).

Environmental Law Centre, 'Multiple chemical sensitivity (MCS) conference', Environmental Law Centre web site (http://www.elc.org.uk).

Equality Authority (Ireland), Dublin, 'Equality in a diverse Ireland', Dublin (http://www.equality.ie).

Estlander, T. and Jolanki, R. (2003), 'Sukupuoli ja ammatti-ihotaudit' ('Gender and occupational skin diseases'), in Luoto, R., Viisainen, K. and Kulmala, I. (eds) *Sukupuoli ja terveys* (*Gender and health*), Vastapaino.

Estola-Partanen, M. (2000), *Muscular tension and tinnitus. An experimental trial of trigger point injections on tinnitus*, Academic dissertation, University of Tampere, Medical School, Tampere University Hospital, Department of Otorhinolaryngology, Vammala, Vammalan Kirjapaino Oy (http://acta.uta.fi/pdf/951–44–4972-X.pdf).

European Agency for Safety and Health at Work (2001a), *How to reduce workplace accidents*, Office for Official Publications of the European Communities, Luxembourg.

European Agency for Safety and Health at Work (2001b), *Health and safety campaigning: getting the message across*, Office for Official Publications of the European Communities, Luxembourg.

European Agency for Safety and Health at Work (2001c), *Preventing work-related slips, trips and falls: factsheet 14*, Office for Official Publications of the European Communities, Luxembourg.

European Agency for Safety and Health at Work (2002a), *How to tackle psychosocial issues and reduce work-related stress,* Office for Official Publications of the European Communities, Luxembourg.

European Agency for Safety and Health at Work (2002b), *Preventing psychosocial risks and stress at work in practice,* Office for Official Publications of the European Communities, Luxembourg.

European Agency for Safety and Health at Work (2002c), *Violence at work: factsheet,* Office for Official Publications of the European Communities, Luxembourg.

European Agency for Safety and Health at Work (2002d), *Bullying at work: factsheet*, Office for Official Publications of the European Communities, Luxembourg.

European Agency for Safety and Health at Work (2002e), *Working on stress,* Magazine 5, Office for Official Publications of the European Communities, Luxembourg.

European Agency for Safety and Health at Work (2002f), *Research on changing world of work,* Working paper, Office for Official Publications of the European Communities, Luxembourg.

European Agency for Safety and Health at Work (2003a), *Promoting health and safety in European small and medium-sized enterprises (SMEs): SME funding scheme 2001–02*, Office for Official Publications of the European Communities (http://agency.osha.eu.int/publications/ other/20030408/en/sme.pdf).

European Agency for Safety and Health at Work (2003b), *SME funding scheme 2002: summary of awarded projects* (http://agency.osha.eu.int/sme/index_en.htm).

European Agency for Safety and Health at Work (2003c), *Improving occupational safety and health in SMEs: examples of effective assistance*, Office for Official Publications of the European Communities, Luxembourg.

European Commission (1991), Commission recommendation of 27 November 1991 on the protection of the dignity of women and men at work, including the code of practice to combat sexual harassment (92/131/EEC)', OJ L 49, 24.2.1992, Luxembourg, pp. 1–8.

European Commission (1997), *Partnership for a new organisation of work*, Green Paper, European Commission, Brussels.

European Commission, Employment and Social Affairs (1998), *Sexual harassment at the workplace in the European Union,* Office for Official Publications of the European Communities, Luxembourg.

European Commission (1999), 'Community strategy for endocrine disrupters: a range of substances suspected of interfering with the hormone systems of humans and wildlife', Communication from the Commission to the Council and the European Parliament, COM(1999) 706 final, 17.12.1999, Brussels.

European Commission (2000a), 'On the guidelines on the assessment of the chemical, physical and biological agents and industrial processes considered hazardous for the safety or health of pregnant workers and workers who have recently given birth or are breastfeeding (Council Directive 92/85/EEC)', Communication from the Commission, COM(2000) 466 final/2, Brussels.

European Commission, Employment and Social Affairs (2000b), *Guidance on work-related stress — Spice of life or kiss of death?*, Office for Official Publications of the European Communities, Luxembourg.

European Commission (2001), *Promoting a European framework for corporate social responsibility*, Green Paper, European Commission, Brussels.

European Commission, Employment and Social Affairs (2001), *Towards a Community strategy* on gender equality (2001–05): equality between men and women, Office for Official Publications of the European Communities, Luxembourg.

European Commission, Employment and Social Affairs, Social Agenda 2, July 2002 (http://europa.eu.int/comm/employment_social/soc_agenda_en.html).

European Commission, Employment and Social Affairs (2002a), 'Sexual harassment outlawed', Social Agenda 2, July 2002, Office for Official Publications of the European Communities, Luxembourg.

European Commission, Employment and Social Affairs (2002b), 'Dialogue between employers and unions enters a new phase', Social Agenda 3, 2002, Office for Official Publications of the European Communities, Luxembourg (http://europa.eu.int/comm/employment_social/social_agenda3_en.pdf).

European Commission (2002c), 'Adapting to change in work and society: a new Community strategy on health and safety at work 2002–06', Communication from the Commission, COM(2002) 118 final, 11.3.2002, Brussels.

European Commission (2002d), Report requested by the Stockholm European Council: 'Increasing labour force participation and promoting active ageing', Report from the Commission to the Council, the European Parliament, the European Economic and Social Committee and the Committee of the Regions, COM(2002) 9 final, 24.1.2002, Brussels (http://europa.eu.int/comm/employment_social/news/2002/feb/com_2002_9_en.pdf).

European Commission, Employment and Social Affairs web site (a), 'European campaign to raise awareness of violence against women' (http://europa.eu.int/comm/employment_social/equ_opp/violence_en.html).

European Commission, Employment and Social Affairs web site (b), 'Teleworking agreement' (http://europa.eu.int/comm/employment_social/news/2002/jul/telework_en.pdf).

European Commission (no date), 'A guide to gender impact assessment', European Commission web site (http://europa.eu.int/comm/employment_social/equ_opp/gender/gender_en.pdf).

European Commission web site, 'Gender equality legislation' (http://europa.eu.int/comm/employment_social/equ_opp/rights_en.html#other).

European Commission gender equality web site, 'Statistics on gender issues' (http://europa.eu.int/comm/employment_social/equ_opp/statistics_en.html).

European Commission gender equality web site, 'Statistics on gender issues — gender segregation in the labour market' (http://europa.eu.int/comm/employment_social/equ_opp/statistics/segregation.pdf).

European Commission gender equality web site, 'Statistics on gender issues — part-time employment' (http://europa.eu.int/comm/employment_social/equ_opp/statistics/parttime.pdf).

European Council (2000), 'Presidency Council conclusions. Lisbon European Council, 23 and 24 March 2000', News release, European Council web site (http://ue.eu.int/Newsroom/LoadDoc.asp?BID=76 DID=60917 from = LANG=1).

European Council (2002), Council Decision 2002/177/EC of 18 February 2002 on guidelines for Member States' employment policies for the year 2002, OJ L 60, 1.3.2002, Office for Official Publications of the European Communities, Luxembourg, pp. 60–69.

European Foundation for the Improvement of Living and Working Conditions (2002a), Quality of work and employment in Europe. Issues and challenges, Foundation Paper No 1, February 2002, Office for Official Publications of the European Communities, Luxembourg (http://www.eurofound.ie/publications/files/EF0212EN.pdf).

European Foundation for the Improvement of Living and Working Conditions (2002b), *Access to employment for vulnerable groups*, Foundation Paper No 2, June 2002, Office for Official Publications of the European Communities, Luxembourg (http://www.eurofound.ie/publications/files/EF0244EN.pdf).

European Foundation for the Improvement of Living and Working Conditions (2002c), *Quality of women's work and employment — Tools for change*, Foundation Paper No 3, December 2002, Office for Official Publications of the European Communities, Luxembourg.

European Foundation for the Improvement of Living and Working Conditions, 'Full-time or part-time work: realities and options. Employment options of the future' (http://www.eurofound.ie/publications/files/EF0021EN.pdf).

European Heart Network (1998), *Social factors, work, stress and cardiovascular disease prevention*, European Heart Network, Brussels.

European Industrial Relations Observatory on-line (1998), 'French national action plan (NAP) on employment adopted', European Foundation for the Improvement of Living and Working Conditions web site (http://www.eiro.eurofound.ie/1998/05/feature/FR9805107F.html).

European Outcome of Depression International Network web site (http://www.markwalton.net/links.asp).

European Parliament, Committee on Employment and Social Affairs (2002), Report on the Commission communication 'Adapting to change in work and society: a new Community strategy on health and safety at work 2002–06' (COM(2002) 118 — C5-0261/2002 — 2002/2124(COS)),16.9.2002, FINAL A5–0310/2002 (http://www2.europarl.eu.int/omk/sipade2?L=EN&OBJID=3904&LEVEL=3&MODE=SIP&NAV=X&LSTDOC=N#Content5c0840).

European Platform for Vocational Rehabilitation (2002), *Participation of women in vocational rehabilitation programmes: a comparative research*, Brussels (http://www.epvr.org/Women_finale.doc).

European Women's Lobby (2000), *Women's health*, European Women's Lobby, Brussels (http://www.womenlobby.org/Document.asp?DocID=94&tod=163423).

European Women's Lobby (2002), *Strengthening the gender dimension in the new Community strategy on health and safety at work 2002–06*, European Women's Lobby, Brussels (http://www.womenlobby.org/Document.asp?DocID=474&tod=161133).

Eurostat, 'Population and social conditions 1999', *Average EU woman earns a quarter less than a man*, Office for Official Publications of the European Communities, Luxembourg.

Eurostat (2002a), 'People in Europe', *Eurostat yearbook. The statistical guide to Europe — Data 1990–2000*, Office for Official Publications of the European Communities, Luxembourg.

Eurostat (2002b), *The life of women and men in Europe: a statistical portrait — Data 1980–2000*, European Commission, Office for Official Publications of the European Communities, Luxembourg.

Eurostat NewCronos, 'Population and social conditions 1999', Tilastokeskus, Eurostat Data Shop, Helsinki.

EWHNET (European Women's Health Network Working Group on Occupational Health)

(2001), 'Gender sensitivity in occupational health', Presented at 'A seminar on the gender dimension in health and safety', 16 November, Belgium EU Presidency/Federal Ministry of Labour and Employment Equal Opportunities Unit with the TUTB, Brussels.

Fagan, C. and Burchell, B. (2002), *Gender, jobs and working conditions in the European Union*, European Foundation for the Improvement of Living and Working Conditions, Office for Official Publications of the European Communities, Luxembourg.

Fagan, C., Warren, T. and McAllister, I. (2001), *Gender, employment and working time preferences in Europe*, European Foundation for the Improvement of Living and Working Conditions, Office for Official Publications of the European Communities, Luxembourg.

Fernandez, M., Schwarz, G. E. and Bell, I. R. (1999), 'Subjective ratings of odorants by women with chemical sensitivity', *Toxicology and Industrial Health* 15(6), pp. 577–581.

Figà-Talamanca, I. (1999), 'Reproductive health and occupational hazards among women workers', in Kane, P. (ed.), *Women and occupational health: issues and policy paper prepared for the Global Commission on Women's Health*, WHO, Geneva.

Figà-Talamanca, I. (2000), 'Reproductive problems among women healthcare workers: epidemiologic evidence and preventive strategies', *Epidemiologic Reviews* 22, pp. 249–260.

Figà-Talamanca, I. et al. (2000), 'Stressful work conditions and menstrual dysfunctions among female hospital employees', *Journal of Healthcare Safety Compliance and Infection Control* 4, pp. 69–74.

Finnish Ministry of Social Affairs, Occupational Safety and Health Act 738/2002: unofficial translation 20.2.2003, Helsinki (http://fi.osha.eu.int/legislation/oshact.pdf).

FIOH, 'Legal provisions on the protection of pregnant women at work', *Työterveiset*, Newsletter of the Finnish Institute of Occupational Health, Special Issue 2/1999.

Forastieri, V. (2000), *Safe work: information note on women workers and gender issues on occupational safety and health*, International Labour Office, Geneva, pp. 1–15.

Forss, S., Karisalmi, S. and Tuuli, P. (2001), *Työyhteisö, jaksaminen ja eläkeajatukset* ('Workplace, resources and retirement thoughts'), *Työssäpysymiseen liittyvän tutkimusprojektin loppuraportti. Eläketurvakeskus*, Raportteja 2001, Helsinki.

Franco, A. and Winqvist, K. (2002), 'Women and men reconciling work and family life', *Statistics in Focus*, Theme 3: Population and social conditions, 9/2002, Eurostat, Office for Official Publications of the European Communities, Luxembourg.

Frankenhaeuser, M., Lundberg, U., Fredrikson, M., Melin, B., Tuomisto, M., Myrsten, A.-L., Hedman, M., Bergman-Losman, B. and Wallin, L. (1989), 'Stress on and off the job as related to sex and occupational status in white-collar workers', *Journal of Organisational Behaviour* 10, pp. 321–346.

Fransson, S., Johansson, L. and Svenaeus, L. (2001), *Highlighting pay differentials between women and men*, Swedish Presidency study, European Commission web site (http://europa.eu.int/comm/employment_social/equ_opp/documents/paydiff_en.pdf).

Fremont, J. (2001), 'Agriculture in Europe: the spotlight on women', *Statistics in Focus*, Theme 5: Agriculture and fisheries, 7/2001, Eurostat, Office for Official Publications of the European Communities, Luxembourg.

Galinsky, T., Waters, T. and Malit, B. (2001), 'Overexertion injuries in home healthcare workers and the need for ergonomics', *Home Health Care Services Quarterly* 20(3), Haworth Press, United States, pp. 57–73.

Garcia, A. and Lega, H. (1999), *The 'second sex' of European trade unionism: research into women and decision-making in trade union organisations*, ETUC, Brussels.

Gates, G. A., Couropmitree, N. N. and Meyers, R. H. (1999), 'Genetic associations in age-related hearing thresholds', *Archives of Otolaryngology — Head and Neck Surgery* 125, pp. 654–659.

Geary, K. G., Irvine, D. and Croft, A. M. (2002), 'Does military service damage females? An analysis of medical discharge data in the British armed forces', *Occupational Medicine* 52(2), pp. 85–90.

Giezen, A. van der (1998), *Women more often in the disability system?* (in Dutch), Elsevier, The Hague.

Giezen, A. van der (2000), *Women, working conditions and disability* (in Dutch), Amsterdam.

Gjesdal, S. and Bratberg, E. (2002), 'The role of gender in long-term sickness and transition to permanent disability benefits. Results from a multi-register-based prospective study in Norway 1990–95', *European Journal of Public Health* 12, pp. 180–186.

Glasgold, A. and Altmann, F. (1966), 'The effect of stapes surgery on tinnitus in otosclerosis', *Laryngoscope* 76, pp. 1624–1632.

Glasgow Healthy City Project, Women's Health Working Group (1997), *Action for women's health. Making changes through organisations*, Glasgow Healthy City Project, Glasgow City Council, Glasgow.

Goldman, M. and Hatch, M. (eds) (2000), *Women and health*, Academic Press, San Diego, San Francisco, New York, Boston, London, Sydney, Tokyo.

Gonäs, L. (2002), 'Gender segregation in the labour market', in Skiöld, L. (ed.), *Women and work — Seminar report*, Worklife and EU enlargement WLE report 4, Swedish National Labour Market Board (AMS) (www.ams.se/wle).

Goudswaard, A. and Andries, F. (2002), *Employment status and working conditions*, European Foundation for the Improvement of Living and Working Conditions, Office for Official Publications of the European Communities, Luxembourg.

Goudswaard, A. and de Nanteuil, M. (2000), *Flexibility and working conditions. A qualitative and comparative study in seven EU Member States*, European Foundation for the Improvement of Living and Working Conditions, Office for Official Publications of the European Communities, Luxembourg.

Goudswaard, A., André, J. C., Ekstedt, E., Huuhtanen, P., Kuhn, K., Peirens, K., OpdeBeeck, R. and Brown, R. (2002), *New forms of contractual relationships and the implications for occupational safety and health*, European Agency for Safety and Health at Work, Office for Official Publications of the European Communities, Luxembourg.

Goudswaard, A., Houtman, I., Kraan, K. and van den Berg, R. (1999), *Working conditions of flexible workers and part-timers*, TNO Work and Employment, Hoofddorp.

Graham, P. J. (2000), 'Improving access to occupational health services', *Occupational Health for Europeans*, Proceedings of the International Symposium, Helsinki, 3 to 5 November 1999, pp.113–117.

Grönkvist, L. and Lagerlöf, E. (1999), 'Work and health among European women', in Messing,

K. (ed.), *Integrating gender in ergonomic analysis*, Trade Union Technical Bureau, Brussels.

Groth, M. V., Burr, H. and Guichard, A. (2000), *Køn, arbejdmijø og helbred. Arbejdsmijø i Danmark 2000*, AMI, Arbejdsmiljøinstituttet.

Gründemann, R. W. M. and Van Vuuren, C. V. (1997), *Preventing absenteeism at the workplace — European research report*, Office for Official Publications of the European Communities, Luxembourg.

Gump, B. B. and Matthews, K. A. (2000), 'Are vacations good for your health? The nine-year mortality experience after the multiple risk factor intervention trial', *Psychosomatic Medicine* 62, pp. 608–612.

Hague, J., Oxborrow, L. and McAtamney, L. (2001), *Musculoskeletal disorders and work organisation in the European clothing industry*, Trade Union Technical Bureau, Brussels.

Hakanen, J. (1999), 'Gender-related differences in burnout', *Työterveiset*, Newsletter of the Finnish Institute of Occupational Health, Special Issue 2/1999, FIOH, pp. 15–17.

Hakola, T. (2002), 'Economic incentives and labour market transitions of the aged Finnish workforce', *Vatt-tutkimuksia* 89, Helsinki.

Hallberg, L. R. and Barrenäs, M. L. (1993), 'Living with a male with noise-induced hearing loss: experiences from the perspective of the spouses', *British Journal of Audiology* 27, pp. 255–262.

Hallberg, L. R. and Carlsson S. G. (1991), 'A qualitative study of the strategies for managing a hearing impairment', *British Journal of Audiology* 25, pp. 201–211.

Hallman, T., Burell, G., Setterlind, S., Oden, A. and Lisspers, J. (2001), 'Psychosocial risk factors for coronary heart disease, their importance compared with other risk factors and gender differences in sensitivity', *Journal of Cardiovascular Risk* 8(1), pp. 39–49.

Hämäläinen, R.-M., Räsänen, K., Husman, K. and Westerholm, P. (2000), 'Survey of the quality and effectiveness of occupational health services in EU Member States, Norway and Switzerland', *Occupational Health for Europeans*, Proceedings of the International Symposium, Helsinki, 3 to 5 November 1999, pp. 129–133.

Hamelsky, S. W., Stewart, W. F. and Lipton, R. B. (2000), 'Epidemiology of headache in women: emphasis on migraine', in Goldman, M. and Hatch, M. (eds), *Women and health*, Academic Press, San Diego, San Francisco, New York, Boston, London, Sydney, Tokyo.

Hansen, J. (2001), 'Increased breast cancer risk among women who work predominantly at night', *Epidemiology* 12, pp. 74–77.

Hardell, L., Hallquist, A., Hansson Mild, K., Påhlson, A. and Lilja, A. (2002), 'Cellular and cordless telephones and the risk of brain tumours', *European Journal of Cancer Prevention* 11(4), pp. 377–386.

Hardell, L., Näsman, A., Pahlson, A., Hallquist, A. and Hansson Mild, K. (1999), 'Use of cellular telephones and the risk of brain tumours: a case-control study', *International Journal of Oncology* 15, pp. 113–116.

Harisinghani, A. (2000), 'Voice problems: an occupational hazard for teachers', Indianest web pages at Boliji.com (http://www.indianest.com/health/articles/01011.htm).

Härmä, M., Kivistö, M., Kalimo, R. and Sallinen, M. (2002), 'Työn vaatimukset, työajat ja uni tietotekniikan ammattilaisilla', *Työn muutos ja hyvinvointi tietoyhteiskunnassa* ('Demands of work, working hours and sleep among the employees of the IT sector', *Change of work and* welfare in the information society), *Sitran raportteja* 22, Helsinki.

Harrison, D. A. and Shaffer, M. A. (1994), 'Comparative examinations of self-reports and perceived absenteeism norms: wading through lake Wobegon', *Journal of Applied Psychology* 79(2), pp. 240–251.

Haslam, C., Brown, S., Hastings, S. and Haslam, R. (2003), *Effects of prescribed medication on performance in the working population*, Health and Safety Executive, HSE Books, Sudbury, UK.

Hatch, M., Figà-Talamanca, I. and Salerno, S. (1999), 'Work stress and menstrual patterns among American and Italian nurses', *Scandinavian Journal of Work and Environmental Health* 25(2), pp. 144–150.

Health and Safety Agency (2001), *Dignity at work: the challenge of workplace bullying — Report of the Task Force on the Prevention of Bullying at Work*, The Stationery Office, Dublin.

Health and Safety Commission, Health Services Advisory Committee (1998), *Manual handling in the health services*, HSE Books, Sudbury, UK.

Health and Safety Commission (1999), *Farmwise: your essential guide to health and safety in agriculture*, HSE Books, Sudbury, UK.

Health and Safety Executive (1997), *Priorities for health and safety in catering activities*, HSE Books, Sudbury, UK (http://www.hse.gov.uk/pubns/cater2.htm).

Health and Safety Executive (2000), *Key messages from the LFS for injury statistics: gender and age, job tenure and part-time working*, HSE Books, Sudbury, UK (http://www.hse.gov.uk/keyart.pdf).

Health and Safety Executive (2002a), *Control of substances hazardous to health regulations 2002 (Fourth edition) — Approved code of practice and guidance L5*, HSE Books, Sudbury, UK.

Health and Safety Executive (2002b), *Preventing dermatitis at work: advice for employers and employees*, HSE Books, Sudbury, UK.

Health and Safety Executive (2003a), 'Preventing slip and trips in education', Press release, 10 April, Health and Safety Executive, London.

Health and Safety Executive (2003b), 'Preventing slip and trip incidents in the education sector', Education information sheet No 2, HSE Books, Sudbury, UK.

Health and Safety Executive (2003c), *Caring for cleaners. Guidance and case studies on how to prevent musculoskeletal disorders*, HSG234, HSE Books, Sudbury, UK.

Hearn, J. (2001), 'Men's violence to women, gendered power and "non-violent" institutions', *Gender and violence in the Nordic countries*, Report from a conference in Koge, Denmark, 23 and 24 November.

HELA (Health and Safety Executive/Local Authorities Enforcement Liaison Committee) (2001), *Advice regarding call centre working practices*, LAC94/1(rev) web site (http://www.hse.gov.uk/lau/lacs/94–1.htm#specific).

Heller, M. and Bergman, M. (1953), 'Tinnitus aurium in normally hearing persons', *Annals of Otology, Rhinology and Laryngology* 62, pp. 73–83.

Herbert, C. (1999), *Preventing sexual harassment at work*, International Labour Office, Geneva.

Hétu, R., Getty, L. and Hung, T. Q. (1995), 'Impact of occupational hearing loss on the lives of the workers in occupational medicine', *State-of-the-Art Reviews* 10, Hanlay and Belfus Inc., Philadelphia, p. 3.

Hétu, R., Jones, L. and Getty, L. (1993), 'The impact of acquired hearing loss on intimate relationships: implications for rehabilitations', *Audiology* 32, pp. 363–381.

Hewlett, S. A. (2002), 'Executive women and the myth of having it all', *Harvard Business Review* 80(4), pp. 66–73.

Hietanen, M., Hämäläinen, A.-M. and Husman, T. (2002), 'Hypersensitivity symptoms associated with exposure to cellular telephones: no causal link', *Bioelectromagnetics* 23, pp. 264–270.

Hochschild, A. R. (1983), *The managed heart. Commercialisation of human feeling*, Amazon, United States.

Hochschild, A. R. and Machung, A. (1997), *The second shift. Working parents and the revolution at home*, Amazon, United States.

Holt, V. and Jenkins, J. (2000), 'Endometriosis', in Goldman, M. and Hatch, M. (eds), *Women and health*, Academic Press, San Diego, San Francisco, New York, Boston, London, Sydney, Tokyo.

Hoogendoorn, W. E., Poppel, M. N. M., van Bongers, P. M., Koes, B. W. and Bouter, L. M. (2000), 'Systematic review of psychosocial factors at work and private life as risk factor for back pain', *Spine* 25, pp. 115–125.

House, J. W. and Brackmann, D. E. (1981), 'Tinnitus: surgical treatment', CIBA Foundation symposium 'Tinnitus', London, pp. 204–212.

Houtman, I. and Dhondt, S. (1994), *Weba and Nova-Weba in relation to health and well-being in employees* (in Dutch), Leiden, NIPG.

Houtman, I. and Kompier, M. A. J. (1997), 'Mental health', *Encyclopaedia for Occupational Health and Safety*, International Labour Office, Geneva.

Houtman, I. and Kornitzer, M. (1999), 'The job stress, absenteeism and coronary heart disease European cooperative study (the JACE study) — Design of a multicentre prospective study', *European Journal of Public Health* 9, pp. 52–57.

Houtman, I. and Van den Heuvel, F. (2001), *Coping capacity from women and men in relation to illness and absenteeism: a literature review* (in Dutch), Elsevier, Doetinchem.

Houtman, I., Bosch, C. M., Jettinghoff, K. and van den Berg, R. (2000a), *Work stress in the police force* (in Dutch), TNO Work and Employment, Hoofddorp.

Houtman, I., Broersen, S., de Heus, P., Zuidhof, A. and Meijman, Th. (2000b), 'The epidemiology of fatigue at work', in Houtman, I., Schaufeli, W. B. and Taris, T. (eds), *Mental fatigue and work* (in Dutch), Samsom Alphen a/d Rijn, pp. 37–64.

Houtman, I., Schoemaker, C. G., Blatter, B. M., de Vroome, E. M. M., van den Berg, R., and Bijl, R. V. (2002), *Psychological complaints, interventions and rehabilitation to work; the prognostic study of Invent* (in Dutch), TNO Work and Employment, Hoofddorp.

Humes, L. E. (1984), 'Noise-induced hearing loss as influenced by other agents and by some physical characteristics of the individual', *Journal of the Acoustical Society of America* 76, pp. 1318–1329.

Hunter, S. J., Shaha, S., Flint, D. and Tracy, D. M. (1998), 'Predicting return to work. A long-term follow-up study of railroad workers after low back injuries', *Spine* 21(21), pp. 2319–2328.

Hwang, P. (1999a), 'Are there fathers at the workplace?', Presentation in the seminar 'European diversities: combining work and family in different settings of working life, family life and culture', Helsinki.

Hwang, P. (1999b), 'Work and family — An analysis of company culture and fatherhood', Presentation in the seminar 'European diversities: combining work and family in different settings of working life, family life and culture', Helsinki.

IDICT (Instituto de Desenvolvimento e Inspecção das Condições de Trabalho), *Campanhas (Campaigns)*, IDICT, Lisbon (http://www.idict.gov.pt/).

ILO (no date), *Gender issues in occupational safety and health: PIACT programme*, International Labour Office, Geneva (http://www.ilo.org/public/english/bureau/gender/occupational safety and health/).

ILO/International Labour Conference (1985), Resolution on equal opportunities and equal treatment for men and women in employment, Adopted by the International Labour Conference at its 71st session in June 1985, International Labour Office, Geneva.

ILO (1992), *Report of the 11th session of the ILO/WHO Joint Committee on Occupational Safety and Health*, Geneva, 27 to 29 April, Ref.: JCOH/XI/92, International Labour Office, Geneva.

ILO, Bureau for Gender Equality (2000), *Decent work for women*, International Labour Office, Geneva (http://www.ilo.org/public/english/bureau/gender).

ILO, ICN, PSI, WHO joint programme on workplace violence in the health sector (2002a), *Framework guidelines for addressing workplace violence in the health sector*, International Labour Office, International Council of Nurses, Public Services International, World Health Organisation, International Labour Office, Geneva.

ILO press release (2002b), 'Joint programme launches new initiative against workplace violence in the health sector', Press release ILO/02/49, International Labour Office, Geneva (http://www.ilo.org/public/english/bureau/inf/pr/2002/49.htm).

INAIL (2002), *Women at work*, Rome (http://www.inail.it/multilingua/inglese/pubblicazioni/donneallavoro/sommario.htm).

InfoBASE Europe news service web site, 'Social partners' framework agreement on teleworking', Europe Factsheet 38 (http://www.ibeurope.com/Fact/38telework.htm).

International Agency for Research on Cancer (IARC) web site (http://www.iarc.fr/).

International Transport Workers' Federation (2000), *ITF Women*, Issue 2/2000, International Transport Workers' Federation, London.

International Transport Workers' Federation (2002), *Women transporting the world: an ITF resource book for trade union negotiators in the transport sector*, International Transport Workers' Federation, London.

ISO R 389 (1964), 'Standard reference zero for the calibration of pure tone audiometers', International Organisation for Standardisation, Geneva.

ISO 1999 (1975), 'Acoustics — Assessment of occupational noise exposure for hearing conservation purposes', International Organisation for Standardisation, Geneva.

ISO 4869–3 (1989), 'Acoustics — Hearing protectors simplified method for the measurement of insertion loss of ear-muff type protectors for quality inspection purposes', International Organisation for Standardisation, Geneva.

ISO 1999 (1990), 'Acoustics — Determination of occupational noise exposure and estimation of noise-induced hearing impairment', International Organisation for Standardisation, Geneva.

ISO 4869–1 (1990), 'Acoustics — Hearing protectors subjective method for the measurement of sound attenuation', International Organisation for Standardisation, Geneva.

ISO 9612-2 (1997), 'Acoustics — Guidelines for the measurements and assessment of exposure to noise in the working environment', International Organisation for Standardisation, Geneva.

ISPESL (Istituto superiore per la prevenzione e la sicurezza del lavoro) (2002), 'Mainstreaming the gender dimension into legislation and preventive action', *Mainstreaming OSH into education*, Unpublished report to the European Agency for Safety and Health at Work.

Jaakkola, J. J. K., Verkasalo, P. A. and Jaakkola, N. (2000), 'Plastic interior materials and respiratory health in pre-school children', *American Journal of Public Health* 90, pp. 797–799.

Johansen, C., Boice Jr, J. D., McLaughlin, J. K. and Olsen, J. H. (2001), 'Cellular telephones and cancer — Nationwide cohort study in Denmark', *Journal of the National Cancer Institute* 93, pp. 203–207.

Johnson, J. V., Stewart, W., Hall, E. M., Fredlund, P. and Theorell, T. (1996), 'Long-term psychosocial work environment and cardiovascular mortality among Swedish men', *American Journal of Public Health* 86(3), pp. 324–331.

Jorna, A. and Offers, E. (1991), *Young women, their work, their future* (in Dutch), VUGA, The Hague.

Kaksonen, R., Pyykkö, I., Rosenhall, U., Starck, J., Toppila, E., Kila, S. and Kere, J. (1998), 'Is genetic hearing loss interactive with noise-induced hearing loss? Advances in noise research', in Prasher, D. and Luxon, L. (eds), *Biological effects of noise*, Vol. 1, Whurr Publishers Ltd, London, pp. 59–70.

Kalimo, R. and Toppinen, S. (1997), *Työuupumus suomen työikäisellä väestöllä* (*Workforce's burnout in Finland*), Työterveyslaitos, Helsinki.

Kane, P. (ed.) (1999a), *Women and occupational health: issues and policy paper prepared for the Global Commission on Women's Health*, World Health Organisation, Geneva.

Kane, P. (1999b), 'Migration, workforce and health', in Kane, P. (ed.), *Women and occupational health: issues and policy paper prepared for the Global Commission on Women's Health*, World Health Organisation, Geneva.

Karasek, R. and Theorell, T. (1990), *Healthy work: stress, productivity and the reconstruction of working life*, BasicBooks, London.

Karjalainen, A., Kurppa, K., Martikainen, R., Klaukka, T. and Karjalainen, J. (2001), 'Work is related to a substantial portion of adult-onset asthma incidence in the Finnish population', *American Journal of Respiratory and Critical Care Medicine* 164, pp. 565–568.

Karjalainen, A., Vänskä, J. and Notkola, V. (1999), 'Koetut työperäiset sairaudet ja oireet, ('Occupational diseases and symptoms'), in Paananen, S. (ed.), *Työn vaarat 1999. Koetut työperäiset sairaudet, työtapaturmat ja työväkivaltatapaukset* (*Dangers at work. Experienced occupational diseases, accidents and violence at work*), Työmarkkinat, 2000, Helsinki.

Karlqvist, L. (2000), 'Investigating the gender gap', *Preventing work-related musculoskeletal disorders*, Magazine 3 of the European Agency for Safety and Health at Work, Office for Official Publications of the European Communities, Luxembourg.

Kauppinen, K. (1993), 'OECD Panel Group on Women, Work and Health', Ministry of Social Affairs and Health, Helsinki.

Kauppinen, K. (1999a), 'The working conditions of men and women in the European Union', *Työterveiset*, Newsletter of the Finnish Institute of Occupational Health, Special Issue 2/1999, FIOH, Helsinki.

Kauppinen, K. (1999b), 'Sexual harassment in the workplace', in Kane, P. (ed.), *Women and occupational health: issues and policy paper prepared for the Global Commission on Women's Health*, World Health Organisation, Geneva.

Kauppinen, K. (2001), 'Gender and working conditions in the European Union', in Bildt, C. and Karlqvist, L. (eds), *Women's conditions in working life*, Arbete och hälsa, vetenskaplig skriftserie 2001/17, Arbetslivsinstitutet, Stockholm.

Kauppinen, K. (2003), *Sexual harassment at the workplace. Women's employment and gender equality in Europe*, EU Socrates project (Contract No 71248-CP-2-2000-1-LT- Grundtvig-ADU).

Kauppinen, K. and Kandolin, I. (1998), *Gender and working conditions in the European Union*, European Foundation for the Improvement of Living and Working Conditions, Office for Official Publications of the European Communities, Luxembourg.

Kauppinen, K. and Otala, L. (1999), *Gender equality, work organisation and well-being: equality standards for a good workplace*, Finnish National Work Development Programme/Finnish Institute of Occupational Health, Helsinki.

Kauppinen, K. and Patoluoto, S. (in press), 'Sexual harassment and violence towards policewomen in Finland', in Gruber, J. and Morgan, P. (eds), *In the company of men: rediscovering the links between male domination and sexual harassment*, Northeastern University Press.

Keith, M., Brophy, J., Kirby, P. and Rosskam, E. (2002), *Barefoot research: a worker's manual for organising work security*, International Labour Office, Geneva.

Keltikangas-Järvinen, L. and Ravaja, N. (2002), 'Relationships between hostility and physiological coronary heart disease risk factors in young adults: moderating influence of perceived social support and sociability', *Psychology and Health* 17(2), pp. 173–190.

Kemmlert, K. and Lundholm, L. (2001), 'Slips, trips and falls in different work groups — with reference to age and from a preventive perspective', *Applied Ergonomics* 32(2), pp. 149–153.

Kessler, R. C. and Frank, R. G. (1997), 'The impact of psychiatric disorders on work loss days', *Psychological Medicine* 27, pp. 861–873.

Kessler, R. C., McGonagle, K. A., Zhao, S., Nelson, C. B., Hughes, M., Eshleman, S., Wittchen, H. U. and Kendler, K. S. (1994), 'Lifetime and 12-month prevalence of DSM-III-R psychiatric disorders in the United States. Results from the national comorbidity survey', *Archives of General Psychiatry* 51, pp. 8–19.

Kinnunen, U. (2002), 'Työ, vanhemmuus ja lapset. Teoksessa tieto ja tekniikka. Missä on nainen?' ('Work, parenthood and children. Knowledge and technology. Where are women?'), in Smeds, R., Kauppinen, K., Yrjänheikki, K. and Valtonen, A. (eds), *Tekniikan Akateemisten Liitto TEK*, Lahti, pp. 74–80.

Kinnunen, U., Sallinen, M. and Rönkä, A. (2001), 'Vanhempien työ ja vanhemmuus nuoren kokemana: yhteydet nuoren hyvinvointiin' ('Parents' work and parenthood seen by a young person: connections to a young person's well-being'), *Psykologia* 36(6), pp. 407–418.

Kipen, H. M., Hallman, W., Kelly-McNeil, K. and Fiedler, N. (1995), 'Measuring chemical sensitivity prevalence: a questionnaire for population studies', *American Journal of Public Health* 85, pp. 574–577.

Kivelä, K. and Lahelma, E. (2000), 'Ansiotyön ja perheen yhdistäminen: kaksinkertainen etu vai taakka naisten hyvinvoinnille ja terveydelle' ('Reconciliation of family and work: double benefit or burden to women's welfare and health'), *Sosiaalilääketieteellinen aikakauslehti* 37, pp. 40–52.

Kivimäki, M., Elovainio, M., Vahtera, J., Virtanen, M. and Stansfeld, S. A. (2003), 'Association between organisational inequity and incidence of psychiatric disorders in female employees', *Psychosocial Medicine* 83, pp. 319–326.

Kivimäki, M., Kuisma, P., Virtanen, M. and Elovainio, M. (2001), 'Does shift work lead to poorer health habits? A comparison between women who had always done shift work with those who had never done shift work', *Work and Stress* 15(1), 2001, pp. 3–13.

Kivimäki, M., Leino-Arjas, P., Luukkonen, R., Riihimäki, H., Vahtera, J. and Kirjonen, J. (2002), 'Work stress and risk of cardiovascular mortality: prospective cohort study of industrial employees', *British Medical Journal* 325, p. 857

Kivimäki, M., Vahtera, J., Thomson, L., Griffiths, A., Cox, T. and Pentti, J. (1997), 'Psychosocial factors predicting employee sickness absence during economic decline', *Journal of Applied Psychology* 82, pp. 858–872.

Kivistö, S., Loponen, M. and Kuosma, E. (2001), *Sairauslomalta takaisin töihin. Siirtymävaiheen psykologinen sopimus* (*From sick leave back to work. Psychological agreement of the transitional phase*), Työterveyslaitos, Helsinki.

Kompier, M. A. J. and Cooper, C. (eds) (1999), *Preventing stress, improving productivity: European case studies in the workplace*, Routledge, London and New York.

Kompier, M. A. J. and Marcelissen, F. (1990), *Handbook on work stress* (in Dutch), NIA, Amsterdam.

Kompier, M. A. J., Gründemann, R. W. M., Vink, P. and Smulders, P. G. W. (1996), *Get going! Ten case studies of successful reduction of absenteeism* (in Dutch), Samsom, Alphen a/d Rijn.

Kornitzer, M., de Smet, P., Boulenguez, C., Backer, G., de Ferrario, M., Houtman, I., Ostergren, P.-O., Sans, S. and Wilhelmsen, L. (2002), 'The demand-control model and incidence of coronary heart disease in a multicentre European project: the JACE study', Presentation at the third ICOH-CVD, Düsseldorf, 20 to 22 March.

Kristensen, T. S., Kornitzer, M., Alfredsson, L. and Marmot, M. (1998), *Social factors, work, stress and cardiovascular disease prevention in the European Union*, European Heart Network, July 1998, Belgium.

Kumpulainen, R. (2000), 'Timber and herring: modernisation and mobility in Finnish Lapland and the Western Islands of Scotland', *Research Reports* No 237, Department of Sociology, University of Helsinki, Helsinki, pp. 1770–1970.

Kuper, H., Singh-Manoux, A. and Marmot, M. (2002), 'The role of effort and rewards at work in relation to risk for coronary heart disease within the Whitehall II study', Presentation at the third ICOH-CVD, Düsseldorf, 20 to 22 March.

Lagerlöf, E. (1998), 'Women's occupational health: a European perspective', TUC seminar on women, work and health, 3 November, TUC, London.

Landrine, H. and Klonoff, E. A. (1997), *Discrimination against women. Prevalence, consequences, remedies*, Sage Publications, United States.

Landsbergis, P. A., Cahill, J. and Schnall, P. (1999), 'The impact of lean production and related new systems of work organisation on worker health', *Journal of Occupational Health Psychology* 4(2), pp. 108–130.

Lärarförbundet (2002), TCO, *Marknadsundersökning* (*A market research*), June 2002, Sweden.

Leino, P. I. and Hänninen, V. (1995), 'Psychosocial factors at work in relation to back and limb disorders', *Scandinavian Journal on Work, Environment and Health* 21, pp. 134–142.

Leszcynski, D., Joenvaara, S., Reivinen, J. and Kuokka, R. (2002), 'Non-thermal activation of the hsp27/p38MAPK stress pathway by mobile phone radiation in human endothelial cells: molecular mechanism for cancer- and blood-brain barrier-related effects', *Differentiation* 70, pp. 120–129.

Lewis, C. E., Bucholz, K. K. and Spitznagel, E. (1996), 'Effects of gender and comorbidity on problem drinking in a community sample', *Alcoholism Clinical and Experimental Research* 20, pp. 466–476.

Linares, D. (2003), 'Women and men in agriculture: a statistical look at the family labour force', *Statistics in Focus*, Theme 5: Agriculture and fisheries, 4/2003, Eurostat, Office for Official Publications of the European Communities, Luxembourg.

Lindbohm, M.-L. (1998), 'Recent Nordic research on occupational hazards and pregnancy outcome', *Women at work*, Proceedings of an international expert meeting, Finnish Institute of Occupational Health, Helsinki.

Lindbohm, M.-L. and Taskinen, H. (2000), 'Reproductive hazards in the workplace', in Goldman, M. and Hatch, M. (eds), *Women and health*, Academic Press, San Diego, San Francisco, New York, Boston, London, Sydney, Tokyo.

Lippel, K. (1993), *Le stress au travail: indemnisation des atteites á la santéen doit québécois, canadien et américan*, Éditions Yvon Blais, Cowansville, p. 28.

Lippel, K. (1996), *Workers' compensation and stress: gender and access to compensation*, Proceedings of the International Congress on Women, Work and Health, Barcelona, 17 to 20 April, pp. 82–91.

Lippel, K. and Demers, D. (1996), *L'invisibilité, facteur d'exclusion: les femmes victims de lesion professionnelles*, Department of Legal Sciences, Université du Québec à Montréal, Montreal.

Lloyd, S. (1997), 'The effects of domestic violence on women's employment'. *Law and Policy* 19(2), pp. 139–167.

Lundberg, U. and Frankenhaeuser, M. (1999), 'Stress and workload of men and women in high-ranking positions', *Journal of Occupational Health Psychology* 4(2), pp. 142–151.

Lutz, H. (undated), *At your service madame! Domestic servants, past and present: gender, class, ethnicity and profession*, Unpublished research project (http://www.vifu.de/new/areas/migration/projects/lutz.html).

Lynch, J., Krause, N., Kaplan, G. A., Tuomilehto, J. and Salonen, J. T. (1997), 'Workplace conditions, socioeconomic status and risk of mortality and acute myocardial infarction: the Kuopio ischaemic heart disease risk factor study', *American Journal of Public Health* 87(4), pp. 617–622.

MacAlister Elliott and Partners (2002), *The role of women in the fisheries sector: report for the European Commission Fisheries DG*, Tender FISH/2000/01-LOT No 1, Final report 1443/R/03/D, Fisheries DG web site (http://europa.eu.int/comm/fisheries/doc_et_publ/liste_publi/studies/women/index.htm).

Maestro, J., Montero, M. A. and Rivero, M. R. (2000), 'Perfiles formativos y necesidades de formación en materia de preventión de riesgos laborales de los empleados públicos españoles', ('Description and needs for occupational safety and health training among public sector employees'), *Prevención, Trabajo y Salud* 7, pp. 4–15.

Magnavits, N. (2002), 'Workplace violence in female nurses', Workshop WeW01:1 in Bildt, C., Gonäs, L., Karlqvist, L. and Westberg, H. (eds), *Women, work and health — IIIrd international congress — Book of abstracts*, Arbetslivsinstitutet, Stockholm.

Marmot, M., Bosma, H., Hemingway, H., Brunner, E. and Stansfeld, S. (1997), 'Contribution of job control and other risk factors to social variations in coronary heart disease incidence', *The Lancet* 350, pp. 235–239

Marmot, M., Feeney, A., Shipley, M., North, F. and Syme, S. L. (1995), 'Sickness absence as a measure of health status and functioning: from the UK Whitehall II study', *Journal of Epidemiology and Community Health* 49, pp. 124–130.

Maschewsky, W. (2002), 'MCS — Oversensitivity or overexposure?', Multiple Chemical Sensitivity International Conference, 8 April, Environmental Law Centre (http://www.elc.org.uk/Conference_Papers/MCS%20-%20oversensitivity%20or%20overexposure.doc).

Massai, G., Ambiente e Lavoro Toscana (2001), 'The lessons of the Tuscany survey into the health of women workers', Presented at 'A seminar on the gender dimension in health and safety', 16 November, Belgium EU Presidency/ Federal Ministry of Labour and Employment Equal Opportunities Unit with the TUTB, Brussels.

Mastekaasa, A. (2000), 'Parenthood, gender and sickness absence', *Social Science and Medicine* 50, pp. 1827–1842.

Mather, C. and Peterken, C. (1999), 'Health promotion: learning from Europe', *Occupational Health Review*, November/December, pp. 23–26 (http://www.bkk.de/ps/tools/download.php?file=/bkk/content/psfile/downloaddatei/42/health_cir3ea7f5b2cfff5.pdf &name=health_circles.pdf&id=96&nodeid=28).

Matthews, S., Manor, O. and Power, C. (1999), 'Social inequalities in health: Are there gender differences?', *Social Sciences and Medicine* 48, pp. 49–60.

Mausner-Dorsch, H. and Eaton, W. W. (2000), 'Psychosocial work environment and depression: epidemiological assessment of the demand-control model', *American Journal of Public Health* 90(11), pp. 1765–1770.

McCarthy, E. and Scannell, J. (2002), *Women, work and health — An Irish perspective*, Report to the Occupational Health and Safety Institute of Ireland.

McCoy, C. A., Carruth, A. K. and Reed, D. B. (2002), 'Women in agriculture: risks for occupational injury within the context of gendered role', *Journal of Agricultural Safety and Health* 8(1), pp. 37–50.

McShane, D. P., Hyde, M. L. and Alberti, P. W. (1988), 'Tinnitus prevalence in industrial hearing loss compensation claimants', *Clinical Otolaryngology and Allied Sciences* 13, pp. 323–330.

Meding, B. and Torén, K. (2001), 'Occupational skin and respiratory diseases', in Marklund, S. (ed.), *Worklife and health in Sweden 2000*, Arbetslivsinstitutet, Stockholm.

Meier, J. (1997), 'Domestic violence, character and social change in the welfare reform debate', *Law and Policy* 19(2), pp. 205–263.

Mejer, L. and Siermann, C. (2000), 'Income poverty in the European Union: children, gender and poverty gaps', Theme 3: 'Population and social conditions', Eurostat (http://europa.eu.int/comm/employment_social/equ_opp/statistics_en.html).

Menckel, E. (2001), 'Threats, violence and harassment in school and worklife', in Marklund, S. (ed.), *Worklife and health in Sweden 2000*, Arbetslivsinstitutet, Stockholm.

Messing, K. (1998), *One-eyed science: occupational health and women workers*, Temple University Press, Philadelphia.

Messing, K. (ed.) (1999), *Integrating gender in ergonomic analysis: strategies for transforming women's work*, Trade Union Technical Bureau for Health and Safety, Brussels.

Messing, K. (2000), 'Ergonomic studies provide information about occupational exposure differences between women and men', *Journal of the American Medical Women's Association* 55(2), pp. 72–75.

Messing, K. and Stellman, J. (1999), 'Understanding occupational diresae in the workforce', in Kane, P. (ed.), *Women and occupational health: issues and policy paper prepared for the Global Commission on Women's Health*, World Health Organisation, Geneva, pp 23-29.

Messing, K., Chatigny, C. and Courville, J. (1998a), '"Light" and "heavy" work in the housekeeping service of a hospital', *Applied Ergonomics* 29, pp. 451–459.

Messing, K., Dumais, L., Courville, J., Seifert, A. M. and Boucher, M. (1994), 'Evaluation of exposure data from men and women with the same job title', *Journal of Occupational Medicine* 36(8), pp. 913–918.

Messing, K., Punnett, L., Bond, M., Alexanderson, K., Pyle, J., Zahm, S., Wegman, D., Stock, S. R. and de Grosbois, S. (2003), 'Be the fairest of them all: challenges and recommendations in the treatment of gender in occupational health research', *American Journal of Industrial Medicine* 43(6), pp. 618–629.

Messing, K., Tissot, F., Saurel-Cubizolles, M.-J., Kaminski, M. and Bourgine, M. (1998b), 'Sex as a variable can be a surrogate for some working conditions factors associated with sickness absence', *Journal of Occupational and Environmental Medicine* 40(3), pp. 251–260.

Middleton, N., Gunnell, D., Whitley, E., Dorling, D. and Frankel, S. (2001), 'Secular trends in antidepressant prescribing in the UK, 1975–98', *Journal of Public Health Medicine* 23, pp. 262–267.

Miller, B. A. and Downs, W. R. (2000), 'Violence against women', in Goldman, M. and Hatch, M. (eds), *Women and health*, Academic Press, San Diego, San Francisco, New York, Boston, London, Sydney, Tokyo.

Miller, K., Greyling, M., Cooper, C. L., Lu, L. and Sparks, K. (2000), 'Occupational stress and gender: a cross-cultural study', *Stress Medicine* 16, pp. 271–278.

Mirrlees-Black, C. (1999), 'Domestic violence: findings from a new British crime survey self-completion questionnaire', *Home Office Research Study* 191, London.

Mohren, D. (2003), 'Results from the NWO research programme "Fatigue at work"', Department of Epidemiology, Maastricht University, Reported in Newsletter 14 of the European Agency for Safety and Health at Work, Office for Official Publications of the European Communities, Luxembourg.

Montero, M. A. and Frutos, P. E. (2000), *Encuesta sobre necesidades formativas de los empleados públicos Europeos en salud laboral* (Survey on occupational safety and health training needs of public sector employees), Leonardo da Vinci project E/97/2/00428/EA/III.2.A/FPC, Federación de Servicios y Administrationes Públicos de Comisiones Obreras, Madrid.

Morata, T. C., Dunn, D. E., Kretshmer, L. W., Lemasters, G. K. and Keith, R. W. (1991), 'Effects of occupational exposure to organic solvents and noise on hearing', *Scandinavian Journal of Work, Environment and Health* 19(4), pp. 245–254.

Mortola, J. (2000), 'Premenstrual syndrome', in Goldman, M. and Hatch, M. (eds), *Women and health*, Academic Press, San Diego, San Francisco, New York, Boston, London, Sydney, Tokyo.

Myers, E. N. and Bernstein, J. M. (1965), 'Salicylate oto-toxicity', *Archives of Otolaryngology* 82, pp. 483–493.

Naber, P. (1991), *Friendship amongst young women*, Acco/Library Youth and Society, Amersfoort.

Naegele, G. (1999), *Active strategies for an ageing workforce*, Conference report, European Foundation for the Improvement of Living and Working Conditions, Office for Official Publications of the European Communities, Luxembourg.

National Centre for Women and Policing web site (http://www.womenandpolicing.org/aboutus.asp).

Naumanen, P. (2002), 'Koulutuksella kilpailukykyä. Koulutuksen yhteys miesten ja naisten työllisyyteen ja työn sisältöön' ('Competitiveness by education. Education's connection to men's and women's employment and content of work'), *Koulutussosiologian tutkimuskeskuksen raportti* 57, Turun yliopisto.

NCHS (1968), 'Hearing status and ear examination: findings among adults, United States 1960–62', *Vital and Health Statistics*, Series 11, No 32, National Centre for Health Statistics, US Department of Health Education and Welfare, Washington, DC.

Nelson, D. L. and Burke, R. J. (2000), 'Women, work stress and health', in Davidson, M. J. and Burke, R. J. (eds), *Women in management. Occupational stress and black and ethnic minority issues*, Part III, Sage Publications, London.

Niedhammer, I., Bugel, I., Goldberg, M., Leclerc, A. and Guéguen, A. (1998), 'Psychosocial factors at work and sickness absence in the Gazel cohort: a prospective study', *Occupational and Environmental Medicine* 55, pp. 735–741.

Niedhammer, I., Saurel-Cubizolles, M.-J., Piciotti, M. and Bonenfant, S. (2000), 'How is sex considered in recent epidemiological publications on occupational risks?', *Occupational and Environmental Medicine* 57, pp. 521–527.

NIOSH (1997), *The effects of workplace hazards on male reproductive health*, National Institute for Occupational Safety and Health, Cincinnati.

NIOSH (1999a), *Providing safety and health protection for a diverse construction workforce: issues and ideas*, Publication No 99–140, National Institute for Occupational Safety and Health, Cincinnati.

NIOSH (1999b), *The effects of workplace hazards on female reproductive health*, National Institute for Occupational Safety and Health, Cincinnati.

NIOSH factsheet (2001), 'Women's health and safety issues at work', Publication No 2001-123, National Institute for Occupational Safety and Health, Cincinnati.

NIOSH web site (a), *Reproductive health*, NIOSH web pages (http://www.cdc.gov/niosh/repropg.html), National Institute for Occupational Safety and Health, Cincinnati.

NIOSH web site (b), *Occupational heart disease*, NIOSH web pages (http://www.cdc.gov/niosh/heartdis.html), National Institute for Occupational Safety and Health, Cincinnati.

North, F. M., Syme, L., Feeney, A., Shipley, M. and Marmot, M. (1996), 'Psychosocial work environment and sickness absence among British civil servants: the Whitehall study', *American Journal of Public Health* 86, pp. 332–340.

Notkola, V. and Vänskä, J. (1999), 'Määräaikaistyöntekijöiden työtapaturmat ja koetut työperäiset sairaudet' ('Occupational accidents and diseases among the temporary employed'), in Paananen, S. (ed.), *Työn vaarat 1999. Koetut työperäiset sairaudet ja työväki-*

valtatapaukset (*Dangers at work. Experienced occupational diseases and violence at work*), *Työmarkkinat*, 2000:15, Tilastokeskus, Helsinki.

Notkola, V. and Virtanen, S. (1998), 'Occupational mortality differences among women and the role of work', in *Women at work*, Proceedings of an international expert meeting, Finnish Institute of Occupational Health, Helsinki.

Nuutinen, I., Kauppinen, K. and Kandolin, I. (1999), *Tasa-arvo poliisitoimessa — työyhteisöjen ja henkilöstön hyvinvointi* (*Equality in the police force — well-being of work organisation and personnel*), Työterveyslaitos, Sisäasiainministeriö, Helsinki.

O'Brien, M. and Shemilt, I. (2003), *Working fathers: earning and caring*, Equal Opportunities Commission, Manchester.

OECD (2001), *Childcare coverage*, OECD employment outlook, reproduced on the European Commission web site (http://europa.eu.int/comm/employment_social/equ_opp/statistics/childcare.pdf).

OECD (2002), *Babies and bosses: reconciling work and family life — Volume 1 — Australia, Denmark and the Netherlands*, OECD, Paris.

Oeij, P. and Wiezer, N. (2002), *New work organisations, working conditions and quality of work: towards the flexible firm?*, European Foundation for the Improvement of Living and Working Conditions, Office for Official Publications of the European Communities, Luxembourg.

O'Hara, R. (2002), *Scoping exercise for research into the health and safety of homeworkers*, Health and Safety Laboratory, HSL/2002/18, Sheffield.

Olgiati, E. and Shapiro, G. (2002), *Promoting gender equality in the workplace*, European Foundation for the Improvement of Living and Working Conditions, Office for Official Publications of the European Communities, Luxembourg.

O'Neill, R. (2000), *Europe under strain. A report of trade union initiatives to combat workplace musculoskeletal disorders*, Trade Union Technical Bureau, Brussels.

Östlin, P. (2002), 'Gender perspective on socio-economic inequalities in health', in Mackenbach, J. and Bakker, M. (eds), *Reducing inequalities in health: a European perspective*, London and New York.

Paoli, P. (2002), *Working conditions in the candidate countries*, European Foundation for the Improvement of Living and Working Conditions, Office for Official Publications of the European Communities, Luxembourg.

Paoli, P. and Merllié, D. (2001), *Third European survey on working conditions 2000*, European Foundation for the Improvement of Living and Working Conditions, Office for Official Publications of the European Communities, Luxembourg (http://www.eurofound.ie/publications/files/EF0121EN.pdf).

Papaleo, B. (2002), 'Presentation of the national ISPESL project to raise awareness of occupational risks to women', Workshop on women, work, safety and health, 15 March, ISPESL (Istituto superiore per la prevenzione e la sicurezza del lavoro), Department of Documentation, Rome.

Papaleo, B., Palmi, S. and Pera, A. (2003), 'Women, health and work: new strategy for research', Unpublished communication, ISPESL (Istituto superiore per la prevenzione e la sicurezza del lavoro), Department of Occupational Medicine, Rome.

Paul, J. (2003), *Health and safety and the menopause: working through the change*, Trades Union Congress, London (http://www.tuc.org.uk/h_and_s/tuc–6316-f0.cfm).

Phillips, A. and Taylor, B. (1980), 'Sex and skill: notes towards feminist economics', *Feminist Review* No 6, pp. 79–89.

Piirainen, H., Elo, A.-L., Hirvonen, M., Kauppinen, K., Ketola, R., Laitinen, H., Lindström, K., Reijula, K., Riala, R., Viluksela, M. and Virtanen, S. (2000), 'Work and health survey', Finnish Institute of Occupational Health, Helsinki.

Piispa, M. and Saarela, K. L. (1999), 'Työväkivalta' ('Worklife violence'), in Paananen, S. (1999), *Työn vaarat 1999 (Dangers at work)*, Työmarkkinat 2000:15, Helsinki.

Plomp, R. (1986), 'A signal-to-noise ratio model for the speech-reception threshold of the hearing impaired', *Journal of Speech and Hearing Research* 29, pp. 146–154.

Plumridge, L. (1999), 'Sex workers and health', in Kane, P. (ed.), *Women and occupational health: issues and policy paper prepared for the Global Commission on Women's Health*, World Health Organisation, Geneva.

Poelmans, S., Compernolle, T., De Neve, H., Buelens, M. and Rombouts, J. (1999), 'Belgium: a pharmaceutical company', in Kompier, M. A. J. and Cooper, C. (eds), *Preventing stress, improving productivity: European case studies in the workplace*, Routledge, London and New York.

Pretty, G., McCarthy, M. and Catano, M. (1992), 'Psychological environments and burnout: gender considerations within the corporation', *Journal of Organisational Behaviour* 13, pp. 701–711.

Pryce, J., Cox, S. and Cox, T. (submitted), 'The impact of long working hours on libido', Submitted to the *Journal of Family and Marital Therapy*.

Punnett, L. and Herbert, R. (2000), 'Work-related musculoskeletal disorders: Is there a gender differential and, if so, what does it mean?', in Goldman, M. and Hatch, M. (eds), *Women and health*, Academic Press, San Diego, San Francisco, New York, Boston, London, Sydney, Tokyo.

Pyykkö, I., Pekkarinen, J. and Starck, J. (1986), 'Sensory-neural hearing loss in forest workers. An analysis of risk factors', *International Archives of Occupational Environmental Health* 59, pp. 439–454.

Rantalaiho, L., Acker, J., Jónasðóttir Melby, K. and Witt-Brattsröm, E. (2002), 'Women's studies and gender research in Finland', *Publications of the Academy of Finland* 8/02, Helsinki (http://www.aka.fi/engl).

Raphael, J. (1997), 'Welfare reform: prescription for abuse? A report on new research studies documenting the relationship of domestic violence and welfare', *Law and Policy* 19(2), pp. 123–137.

Rea, T. and Buchwald, D. (2000), 'Chronic fatigue syndrome', in Goldman, M. and Hatch, M. (eds), *Women and health*, Academic Press, San Diego, San Francisco, New York, Boston, London, Sydney, Tokyo.

Rees, D. (1993), 'Research note: reliability of self-report sickness absence data in the health service', *Health Services Management Research* 6(2), pp. 140–141.

Reilly, B., Paci, P. and Hall, P. (1995), 'Unions, safety committees and workplace injuries', *British Journal of Industrial Relations* 33(2).

Reproductive Health Outlook web site (http://rho.org/html/aboutrho.htm).

Risking Acoustic Shock Conference (2001), Freemantle, Western Australia, September 2001.

Robins, L. N. and Regier, D. A. (eds) (1991), *Psychiatric disorders in America*, The Free Press, New York.

Rogers, R. and Salvage, J. (1988), *Nurses at risk: a guide to health and safety at work*, Heinemann Professional Publishing, London.

Rohlfs, I., Valls-Llobet, C., Lopez, M., Artazcoz, L. and Cirera, E. (2002), 'The total workload of women and men in a cross-cultural perspective', in Bildt, C., Gonäs, L., Karlqvist, L. and Westberg, H. (eds), *Women, work and health — IIIrd international congress — Book of abstracts*, Arbetslivsinstitutet, Stockholm.

Rosskam, E., *Women work to close the 'occupational safety gender gap'*, International Labour Office web site (http://www.ilo.org/public/english/bureau/gender/publ/rosskam.htm).

Rout, U. R., Cooper, C. and Kerslake, H. (1997), 'Working and non-working mothers: a comparative study', *Women in Management Review* 12(7), pp. 264–275.

Ruben, C. H., Burnett, C. A., Halperin, W. E. and Seligman, P. J. (1993), 'Occupation as a risk identifier for breast cancer', *American Journal of Public Health* 83, pp. 1311–1315.

Rylander, L., Axman, A., Torén, K. and Albin, M. (2002), 'Reproductive outcome among female hairdressers', *Occupational and Environmental Medicine* 59, pp. 517–522.

Sala, E., Airo, E., Olkinuora, P., Laine, A., Suonpää, J. P., Tähtinen, E. and Portnoi, H. (1999), *Päiväkotien kasvatushenkilöiden äänihäiriöt ja äänihäiriöiden työperäiset taustatekijät (Daycare employees' voice problems and their occupational explanatory factors)*, Työsuojelurahaston loppuraportti 97030, Turku.

Salerno, S., Bosco, M. G. and Valcella, F. (2002a), 'Mental health and working environment in a group of public health service users' ('Salute mentale e ambiente di lavoro in un gruppo di utenti dei servizi pubblici di prevenzione'), *La Medicina del Lavoro* 93(4), pp. 329–337.

Salerno, S., Bosco, M. G. and Figà-Talamanca, I. (2002b), 'Stress in Italian working women', Workshop SUW05, in Bildt, C., Gonäs, L., Karlqvist, L. and Westberg, H. (eds), *Women, work and health — IIIrd international congress — Book of abstracts*, Arbetslivsinstitutet, Stockholm.

Sallmén, M. (2000), 'Fertility among workers exposed to solvents or lead. Finnish Institute of Occupational Health', *People and Work Research Reports* 37, Otamedia, Espoo.

Sallmén, M. and Taskinen, H. (1998), 'Finnish studies on fertility among female workers', *Women at work, People and Work Research Reports* 20, Finnish Institute of Occupational Health, Helsinki, pp. 121–125.

Sataloff, R. T., Sataloff, J. and Hawkshaw, M. (1993), *Tinnitus in occupational hearing loss — Second edition*, Sataloff, R. T. and Sataloff, J. (eds), Marcel Dekker Inc., New York.

Saurel-Cubizolles, M.-J., Fougeyrollas-Schwebel, D., Camard, R. and the Enveff group (2002), 'Health and violence at work against women: national data in France', Workshop WeW01:3, in Bildt, C., Gonäs, L., Karlqvist, L. and Westberg, H. (eds), *Women, work and health — IIIrd international congress — Book of abstracts*, Arbetslivsinstitutet, Stockholm.

Savolainen, S. and Lehtomäki, K. (1996), 'Hearing protection in acute acoustic trauma in Finnish conscripts', *Scandinavian Audiology* 25(1), pp. 53–58.

Schaufeli, W. B. and Enzmann, D. (1998), *The burnout companion for research and practice: a critical analysis,* Taylor and Francis, London.

Schernhammer, E. S., Laden, F., Speizer, F. E., Willet, W. C., Hunter, J., Kawachi, I. and Colditz, G. (2001), 'Rotating night shifts and risk of breast cancer in women participating in the nurses' health study', *Journal of the National Cancer Institute* 93(20).

Schie, E. van (1997), 'Career aspirations of men and women before coming into the labour market' (in Dutch), *Gedrag Organisatie* 10(5), pp. 286–299.

Schippers, A. (2002), *The special blight of migrant workers,* Radio Netherlands web site (http://www.rnw.nl/society/html/migrant020102.html).

Sennet, R. (1998), *The corrosion of character. The personal consequences of work in the new capitalism*, Norton, New York, London.

Sjögren, J. and Rappe, T. E. (2002), *Diagnos: duktig. Handbok för överambitiösa tjejer och alla andra som borde bry sig* (Diagnosed competent. A handbook for overambitious girls and everyone else who thinks they should care), Bokförlaget.

Skiöld, L. (ed.) (2002), *Women and work — Seminar report*, Worklife and EU enlargement WLE report 4, Swedish National Labour Market Board (AMS), Stockholm (www.ams.se/wle).

Smith, D., Atkinson, R., Junko, M. and Yarnagata, Z. (2002), 'Occupational skin diseases in nursing', *Australian Nursing Journal* 51, June (http://www.anf.org.au/pdf/0206_clin_update.pdf).

Sochert, R. (1999), *Gesundheitsbericht und Gesundheitszirkel*, BAuA, Dortmund/Berlin.

Spagat, 'Innovative Gesundheitsförderung berufstätiger Frauen', Information provided by ppm forschung + beratung — Projekt Spagat, Spagat project web site (http://www.spagat.or.at/).

Spanish Ministry of Employment and Social Affairs (Ministerio de Trabajo y Asuntos Sociales) (2002), *Proyecto Europeo para la mejora de la seguridad y la salud laboral por rezones de reproducción y maternidad* (European project for the improvement of security and health in the workplace due to reproduction pregnancy reasons), Madrid.

Stansfeld, S. (2002), 'Work, personality and mental health', *British Journal of Psychiatry* 181, pp. 96–98.

Steele, M., Alexander, M., Stephen, K. and Duffin, L. (1997), *Changes at work: the 1995 Australian workplace industrial relations survey*, Addison Wesley Longman, Melbourne.

Steenland, K., Johnson, J. and Nowlin, S. (1997), 'A follow-up study and heart disease among males in the NHANES1 population', *American Journal of Industrial Medicine* 4(31), pp. 256–260.

Stefanovska, T. and Pidlisnuk, V. (2002), 'Ukraine struggles with pesticides — Women bear the brunt', *Pesticide News* 57, Pesticides Action Network, London.

Stellman, J. M. and Lucas, A. (2000), 'Women's occupational health: international perspectives', in Goldman, M. and Hatch, M. (eds), *Women and health,* Academic Press, San Diego, San Francisco, New York, Boston, London, Sydney, Tokyo.

Stenberg, B. and Wall, S. (1995), 'Why do women report sick-building syndrome more often than men?', *Social Science and Medicine* 40(4), pp. 491–502.

Stronks, K., van de Mheen, H., van den Bos, J. and Mackenbach, J. P. (1995), 'Smaller socio-economic inequalities in health among women: the role of employment status', *International Journal of Epidemiology* 24, pp. 559–568.

Strowbridge, N. F. (2002), 'Musculoskeletal injuries in female soldiers: analysis of cause and type of injury', *Journal of the Royal Army Medical Corps* 148, pp. 256–258.

Sutela, H. (1999), 'Tasa-arvo, oikeudenmukaisuus, työpaikan sosiaaliset suhteet' ('Equality, equity, workplace social relations'), in Lehto, A.-M. and Sutela, H., *Tasa-arvo työoloissa (Equality at work), Työmarkkinat* 19, Tilastokeskus, Helsinki.

Sutela, H., Vänskä, J. and Notkola, V. (2001), 'Pätkätyöt Suomessa 1990-luvulla' ('Temporary employment in Finland in the 1990s'), *Työmarkkinat* 2001:1, Tilastokeskus, Helsinki.

Taris, T., Schaufeli, W. B., Schreurs, P. and Caljé, D. (2000), 'Burned out in education: stress, mental fatigue and absenteeism amongst teachers' (in Dutch), in Houtman, I., Schaufeli, W. B. and Taris, T. (eds), *Mental fatigue and work* (in Dutch) Samsom, Alphen a/d Rijn, pp. 97–106.

Taskinen, H. K., Kyyrönen, P., Sallmén, M., Virtanen, S. V., Liukkonen, T. A., Huida, O., Lindbohm, M. L. and Anttila, A. (1999), 'Reduced fertility among female woodworkers exposed to formaldehyde', *American Journal of Industrial Medicine* 36(1), pp. 206–212.

Taylor, R. (2002), *Managing workplace change*, Economic and Social Research Council, Swindon.

Tiihonen, A. and Koivisto, N. (eds) (2002), *Allowed to care/allowed to intervene: sexual harassment in sports — A guidebook for adults*, Finnish Sports Federation, SLU publications series 10/02, Helsinki.

TUC (forthcoming), 'TUC survey of safety reps 2002', Trades Union Congress, London.

TUTB, *Results of the survey*, Trade Union Technical Bureau web site (http://www.etuc.org/tutb/uk/pdf/survey-document2.pdf).

TUTB (2000), *Trade union initiatives across Europe*, Women, work and health special report, TUTB Newsletter, 13 March 2001, Brussels, pp. 21–23.

UK Department of Trade and Industry (2000), *Creating a work–life balance: a good practice guide for employers*, Department of Trade and Industry Publications, London (http://www.dti.gov.uk/work-lifebalance/docs/cawlbf.doc).

UNI-Europa and EFCI/FENI (2001), *Health and safety in the office cleaning sectors. European manual for employees*, UNI-Europa, Brussels (http://www.union-network.org/uniproperty.nsf/c56501bbc7a97749c1256b3c004f96fe/cf3ccc8d55dce0fcc1256b11002e1fea?OpenDocument).

Unison (1997), *Unison members' experience of workplace bullying*, Unison, London.

Unison (2001), *Women's health and safety: a guide for Unison safety representatives*, Unison, London.

United Nations (1995), *International migration policies and the status of female migrants: Proceedings of the United Nations Expert Group Meeting, San Miniato, Italy, 28 to 31 March 1990*, ST/ESA/SER.R/126, Population Division, United Nations, New York.

Vagg, P. R., Spielberger, C. D. and Wasala, C. F. (2002), 'Effects of organisational level and gender on stress in the workplace', *International Journal of Stress Management* 9(4), pp. 243–261.

Vahtera, J., Kivimäki, M. and Pentti, J. (1997), *Effect of organisational downsizing on health of employees, The Lancet* 350, pp. 1124–1128.

Valls-Llobet, C. (2002), 'Women's health conditions after occupational chemical exposures', Workshop MoW03, in Bildt, C., Gonäs, L., Karlqvist, L. and Westberg, H. (eds), *Women, work and health — IIIrd international congress — Book of abstracts*, Arbetslivsinstitutet, Stockholm.

Valls-Llobet, C., Borra, G., Doyal, L. and Torns, T. (1999), 'Household labour and health', in Kane, P. (ed.), *Women and occupational health: issues* and policy paper prepared for the Global Commission on Women's Health, World Health Organisation, Geneva.

Vartia, M. (2003), *Workplace bullying: a study on the work environment, well-being and health*, People and Work Research Reports 56, Finnish Institute of Occupational Health, Helsinki.

Veerman, T. J., Schoemaker, C. G., Cuelenaere, B. and Bijl, R. V. (2001), *Mental disability* (in Dutch), Elsevier, Doetinchem.

Verbrugge, L. M. (1985), 'Gender and health: an update on hypotheses and evidence', *Journal of Health and Social Behaviour* 26, pp. 156–182.

Verbrugge, L. M (1986), 'From sneezes to adieus: stages of health for American men and women', *Social Sciences Medicine* 22(11), pp. 1195–1212.

Verschenuren, R., de Groot, B. and Nossent, S. (1995), *Working conditions in hospitals in the European Union*, European Foundation for the Improvement of Living and Working Conditions, Office for Official Publications of the European Communities, Luxembourg.

Vickery, G. and Wurzburg, G. (1996), 'Flexible firms, skills and employment', *The OECD Observer* No 202, October/November, OECD, Paris.

Vingård, E. and Hagberg, M. (2001), 'Work factors and musculoskeletal disorders', in Marklund, S. (ed.), *Worklife and health in Sweden 2000*, Arbetslivsinstitutet, Stockholm.

Vinke, H., Andriessen, S., van den Heuvel, S. G., Houtman, I., Rijnders, S., Van Vuuren, C. V. and Wevers, C. W. J. (1999), *Women and rehabilitation for work* (in Dutch), TNO Work and Employment, Hoofddorp.

Vinke, H. and Wevers, C. (1999), 'Women, disability and rehabilitation in work' (in Dutch), in Houtman, I., Smulders, P. and Klein Hesselink,

D. J. (eds), *Trends in work*, Samsom Alphen a/d Rijn, pp. 217–236.

Virtanen, M., Kivimäki, M., Elovainio, M., Vahtera, J. and Cooper, C. (2001), 'Contingent employment, health and sickness absence', *Scandinavian Journal of Work, Environment and Health* 27, pp. 365–372.

Vogel, L. (1999), *Occupational health in central government administration: a comparative study of the implementation of selected provisions of the framework directive in Austria, Spain, France and the United Kingdom*, Trade Union Technical Bureau, Brussels.

Vogel, L. (2002), 'The gender dimension in health and safety. Initial findings of a European survey', TUTB Newsletter 18, Trade Union Technical Bureau, Brussels, pp. 13–17 (http://www.etuc.org/tutb/uk/survey.html).

Voice Care Network, *More care for your voice*, Kenilworth, UK (www.voicecare.org.uk).

Wamala, S.-P., Mittleman, M. A., Horsten, M., Schenck-Gustafsson, K. and Orth-Gomer, K. (2000), 'Job stress and the occupational gradient in coronary heart disease risk in women. The Stockholm female coronary risk study', *Social Science and Medicine* 51(4), pp. 481–489.

WAO, WAZ and Wajong (2001), *Developments in the use of disability pensions in the Netherlands in 2000*, Landelijk Instituut Sociale Verzekeringen (LISV), Amsterdam.

Webster, J. and Schnabel, A. (1999), *Participating on equal terms*, European Foundation for the Improvement of Living and Working Conditions, Office for Official Publications of the European Communities, Luxembourg.

Wedderburn, A. (ed.) (2000), *Shift work and health*, European Foundation for the Improvement of Living and Working Conditions, Office for Official Publications of the European Communities, Luxembourg.

Weinberg, A. and Creed, F. (2000), 'Stress and psychiatric disorder in healthcare professionals and hospital staff', *The Lancet* 355, pp. 533–537.

Westberg, H. (1998), 'Where are women in today's workplace?', in Kilbom, Å., Messing, K. and Bildt Thorbjörnsson, C. (eds), *Women's health at work*, Arbetslivsinstitutet, Sweden.

Westerholm, P. (1998), 'The heart — a weak spot. Gender, work and cardiovascular disease', in Kilbom, Å., Messing, K. and Bildt Thorbjörnsson, C. (eds), *Women's health at work*, Arbetslivsinstitutet, Sweden.

Westman, M. and Etzion, D. (2001), 'The impact of vacation and job stress on burnout and absenteeism', *Psychology and Health* 16, pp. 595–606.

Whatmore, L., Cartwright, S. and Cooper, C. (1999), 'United Kingdom: evaluation of a stress-management programme in the public sector', in Kompier, M. A. J. and Cooper, C. (eds), *Preventing stress, improving productivity: European case studies in the workplace*, Routledge, London and New York.

White, B. (2000), 'Lessons from the careers of successful women', in Davidson, M. J. and Burke, R. J. (eds), *Women in management*, Part II, Sage Publications, London.

WHO (undated), *Gender disparities and mental health*, World Health Organisation web site, 'Gender and women's mental health' (http://www5.who.int/mental_health/main.cfm?p=0000000022), Geneva.

WHO (1980), *International classification of impairments, disabilities and handicaps*, World Health Organisation, Geneva

WHO (1994a), *A global strategy on occupational health for all*, World Health Organisation web site (http://www.who.int/oeh/OCHweb/OCHweb/OSHpages/GlobalStrategy/GlobalStrategy.htm). Geneva.

WHO (1994b), *Monica project (monitoring cardiovascular disease)*, World Health Organisation, Geneva.

WHO (1998), Gender and health technical paper, World Health Organisation web site (http://www.who.int/frh-whd/GandH/GHreport/gendertech.htm), Geneva.

WHO (2001), 'Mainstreaming gender equity in health: the need to move forward', Madrid statement, Gender mainstreaming health policies in Europe, gender mainstreaming programme, Division of Technical Support, World Health Organisation, Madrid.

WHO (2002a), *Integrating gender perspectives into the work of the World Health Organisation: WHO gender policy*, World Health Organisation web site (http://whqlibdoc.who.int/hq/2002/a78322.pdf), Geneva.

WHO (2002b), 'Selected occupational risks', *World Health Report 2002*, Chapter 4.9, World Health Organisation web site (http://www.who.int/whr/2002/chapter4/en/index8.html), Geneva.

Wickström, G. and Joki, M. (1998), 'Utilisation of official breaks and take-up of short breaks in nursing work', in Lehtinen, S., Taskinen, H. and Rantanen, J. (eds), *Women at work*, Proceedings of an international expert meeting, *People and Work Research Reports* 20, Finnish Institute of Occupational Health, Helsinki.

Williams, C. (2001), 'Health meeting resume', 'Filtration and health — clean air in the workplace', Birmingham, March, The Filtration Society (http://www.lboro.ac.uk/departments/cg/research/filtration/abstracts_health_meeting.htm).

Williams, K. and Umberson, D. (2000), 'Women, stress and health', in Goldman, M. and Hatch, M. (eds), *Women and health*, Academic Press, San Diego, San Francisco, New York, Boston, London, Sydney, Tokyo.

Wise, L. A., Krieger, N., Zierler, S. and Harlow, B. L. (2002), 'Lifetime socioeconomic position in relation to onset of perimenopause', *Journal of Epidemiology and Community Health* 56(11), pp. 851–860.

Wollershein, J. P. (1993), 'Women workers', in Headapohl, D. M. (ed.), *State-of-the-art reviews*, Occupational Medicine.

Woods, V., Buckle, P. and Haisman, M. (1999), *Musculoskeletal health of cleaners*, Health and Safety Executive, Contract research report 215/1999, HSE Books, Sudbury, UK.

WorkCover NSW (2001), *Health and safety guidelines for brothels*, Sydney (http://www.workcover.nsw.gov.au/Publications/pdf/workcover_brothels.pdf).

Wrangsö, K., Wallenhammar, L. M., Ortengren, U., Barregard, L., Andreasson, H., Bjorkner, B., Karlsson, S. and Meding, B. (2001), 'Protective gloves in Swedish dentistry: use and side effects', *British Journal of Dermatology* 145(1), pp. 32–37.

Wynne, R., Clarkin, N., Cox, T. and Griffits, A. (1997), *Guidance on the prevention of violence at work*, European Commission, Office for Official Publications of the European Communities, Luxembourg.

Yeandle, S., Gore, T. and Herrington, A. (1999), *Employment, family and community activities: a new balance for women and men*, European Foundation for the Improvement of Living and Working Conditions, Office for Official Publications of the European Communities, Luxembourg.

Yehuda, R. (2002), 'Post-traumatic stress disorder', *New England Journal of Medicine* 346(2), pp. 108–114.

Ylikoski, M., Lehtinen, S., Kaadu, T. and Rantanen, J. (eds) (2002), *The FinEst bridge — Finnish–Estonian collaboration in occupational health*, Finnish Institute of Occupational Health, Helsinki (http://www.sm.ee/telematic/twinning/finest_bridge.pdf).

York Consulting (2003), *Pilots to explore the effectiveness of workers' safety advisors: interim findings*, Report to the Health and Safety Commission, available at the TUC web site (http://www.tuc.org.uk/h_and_s/tuc–6505-f0.pdf).

Zahm, S. H. (2000), 'Women at work', in Goldman, M. and Hatch, M. (eds), *Women and health*, Academic Press, San Diego, San Francisco, New York, Boston, London, Sydney, Tokyo.

Zahm, S. H., Pottern, L. M., Lewis, R. D., Ward, M. H. and White, D. W. (1994), 'Inclusion of women and minorities in occupational cancer and epidemiological research', *Journal of Occupational Medicine* 36(8), pp. 842–847.

Zahm, S. H., Ward, M. H. and Silverman, D. T. (2000), 'Occupational cancer', in Goldman, M. and Hatch, M. (eds), *Women and health*, Academic Press, San Diego, San Francisco, New York, Boston, London, Sydney, Tokyo.

Zitting, A. and Husgafvel-Pursiainen, K. (2002), 'Health effects of environmental tobacco smoke', *Scandinavian Journal of Work, Environment and Health* 28, Supplement 2.

European Agency for Safety and Health at Work

RESEARCH

ANNEXES

ANNEX 1: ACKNOWLEDGEMENTS

Also, some of the Agency staff members have contributed to the report, including: Markku Aaltonen (Topic Centre manager), Eusebio Rial-González, Christa Sedlatschek, Brenda O'Brien, Mónica Vega and Paola Piccarolo.

Sarah Copsey,
Project Manager

The Agency would like to thank Kaisa Kauppinen and Riitta Kumpulainen, FIOH, Finland, and Irene Houtman, TNO Work and Employment, the Netherlands, for their contribution to this report, and others, also from the Topic Centre on Research — Work and Health, who co-operated in its production: Anneke Goudswaard, TNO Work and Employment, the Netherlands; Maria Castriotta, ISPESL, Italy; Alan Woodside, OSHII, Ireland; Birgit Aust, AMI, Denmark; Veerle Hermans, Prevent, Belgium; Dolores Solé, INSHT, Spain; Karl Kuhn and Ellen Zwink, BAuA, Germany.

The Agency would also like to thank the following who provided valuable comments and suggestions with respect to this report.

The Agency's focal points and other network group members; Carina Bildt, National Institute for Working Life, Sweden; Lesley Doyal, University of Bristol, UK; Marilyn Fingerhut, WHO; Carin Hakansta and colleagues, ILO; Karen Messing, Cinboise, Université du Québec in Montreal, Canada; Naiomi Swanson and colleagues, NIOSH; and Sabrina Tesoka, European Foundation for the Improvement of Living and Working Conditions.

ANNEX 2: PROJECT ORGANISATION, PARTICIPANTS AND EXPERTS

Agency's project manager

Sarah Copsey
European Agency for Safety and Health at Work
Gran Vía, 33
E-48009 Bilbao

Project members of the Topic Centre on Research — Work and Health

Task leaders:

Kaisa Kauppinen and Riitta Kumpulainen
Finnish Institute of Occupational Health (FIOH)
Työterveyslaitos
Topeliuksenkatu 41A
FIN-00250 Helsinki

Irene Houtman and Anneke Goudswaard
TNO Work and Employment
Polarisavenue 151
PO Box 718
2130 AS Hoofddorp
Netherlands

Task members:

Maria Castriotta
National Institute of Occupational Safety and Prevention (ISPESL)
Via Urbana
I-167-00184 Rome

Alan Woodside
Occupational Safety and Health Institute of Ireland (OSHII)
University Road
Galway
Ireland

Birgit Aust
Arbejdsmiljøinstituttet (AMI)
Lersø Parkallé 105
DK-2100 Copenhagen

Veerle Hermans
Prevent
Rue Gachard 88, boîte 4
B-1050 Brussels

Dolores Solé
Instituto Nacional de Seguridad e Higiene en el Trabajo (INSHT)
Dulcet, 2–10
E-08034 Barcelona

Karl Kuhn and Ellen Zwink
Bundesanstalt für Arbeitsschutz und Arbeitsmedizin (BAuA)
Friedrich-Henkel-Weg 1–25
D-44149 Dortmund

Agency's project manager of the Topic Centre on Research — Work and Health

Markku Aaltonen
European Agency for Safety and Health at Work
Gran Vía, 33
E-48009 Bilbao

The leaders of the Topic Centre on Research — Work and Health

Jean-Luc Marié
INRS
30, rue Olivier Noyer
F-75680 Paris Cedex 14

Jean-Claude André
INRS — Centre de Lorraine
Avenue de Bourgogne, BP 27
F-54501 Vandoeuvre Cedex

Participants in the expert meeting, Rome, 15 March 2002

T. Baccolo (ISPESL, Italy)
L. Benedettini (CGIL, Italy)
C. Bildt (NIWL, Sweden)
M. Castriotta (ISPESL, Italy)
A. Ciani Passeri (Prev. Dept. Florence, Italy)
S. Copsey (European Agency for Safety and Health at Work)
G. Costa (ARPA Piemonte, Italy)
G. Delpier (Natl. Comm. Equal. Opport., Italy)
I. Figà-Talamanca (Univ. La Sapienza, Italy)
C. Frascherl (CISL, Italy), G. Galli (UIL, Italy)
P. Germini (Natl. Comm. Equal Opport., Italy)
E. Giuli (ISPESL, Italy)
A. Goudswaard (TNO, the Netherlands)
V. Hermans (Prevent, Belgium)
M. Hiltunen (Min. Soc. Affairs, Finland)
I. Houtman (TNO, the Netherlands)
K. Kauppinen (FIOH, Finland)
K. Kuhn (BAuA, Germany)
B. Papaleo (ISPESL, Italy)
S. Perticaroli (ISPESL, Italy)
S. Salerno (ENEA, Italy)
D. Solé (INSHT, Spain)
E. Volturo (CDF ASL Milan, Italy)
A. Woodside (OSHII, Ireland)
E. Zwink (BAuA, Germany)

Additional assisting experts to the report

Musculoskeletal disorders: H. Riihimäki (FIOH)
Respiratory disorders, asthma and sick-building syndrome: K. Reijula (FIOH)
Skin diseases: R. Jolanki (FIOH) and T. Estlander (FIOH)
Hearing disorders: E. Toppila (FIOH)
Cancer: T. Kauppinen (FIOH); mobile phones: M. Hietanen (FIOH)
Reproductive health: M.-L. Lindbohm (FIOH), H. Taskinen (FIOH) and I. Figà-Talamanca (Univ. La Sapienza)
Absenteeism, disability and rehabilitation: R. Gründemann (TNO),

Also assisted: H. Huuhtanen (FIOH), S. Patoluoto (FIOH) and T. Veriö-Piispanen (FIOH)

ANNEX 3: EXTRACTS RELEVANT TO GENDER FROM THE COMMUNITY STRATEGY ON HEALTH AND SAFETY AT WORK 2002–06

In its Community strategy on health and safety at work 2002–06, the European Commission specifically includes mainstreaming the gender issue into occupational safety and health among its objectives. The relevant text is reproduced below:

'Introduction

It follows that working towards a healthier occupational environment has to be addressed as part of the general trend in economic activities (more service-oriented), forms of employment (more diversified), the active population (more women and an older working population) and society in general (more diverse, but more marked by social exclusion) [...].

2.1. Changes in society

2.1.1. An increasingly feminised society

The growing percentage of women in employment has been evident for decades. It was one of the fundamental objectives laid down in Lisbon, against the background of an ageing working population, and it introduces a new dimension into the subject of health and safety at work.

As many as 83 % of employed women work in services, which explains why they suffer a much lower rate of accidents and occupational illnesses than men, and why they stand much less risk of being involved in an accident. Nonetheless, the trend is not a good one, since the kind of work in which women predominate is generating a growing accident rate, including fatal accidents. Moreover, although women accounted [in 1995] for only 17.8 % of diagnosed occupational illnesses, the proportion was much higher in certain groups: 45 % of allergies, 61 % of infectious illnesses, 55 % of neurological complaints, 48 % of hepatitic and dermatological complaints. These figures underline the importance of gender in terms of occupational illnesses.

Preventive measures and the assessment arrangements, and the rules for awarding compensation must take specific account of the growing proportion of women in the workforce, and of the risks to which women are particularly liable. These measures must be based on research covering the ergonomic aspects, workplace design, and the effects of exposure to physical, chemical and biological agents, and pay heed to the physiological and psychological differences in the way work is organised.

3.1. For a global approach to well-being at work

The objective of the Community's policy on health and safety at work must be to bring about a continuing improvement in well-being at work, a concept which is taken to include the physical, moral and social dimensions. In addition, a number of complementary objectives must be targeted jointly by all the players [...].

2. Mainstreaming the gender dimension into risk evaluation, preventive measures and compensation arrangements, so as to take account of the specific characteristics of women in terms of health and safety at work [...].

3.3.2. Encouraging innovative approaches

1. Benchmarking

The Commission will:

— consider to propose, in 2002, an amendment to the employment guidelines, calling on the Member States to adopt national quantified objectives for reducing accidents at work and occupational illnesses, giving specific attention to sectors with a high accident frequency rate, and mainstreaming the gender and age dimensions; [...].'

Source: European Commission (2002c), Communication from the Commission, 'Adapting to change in work and society: a new Community strategy on health and safety at work 2000–02', COM(2002) 118 final, C5-0261/2002–2002/2124(COS), 11 March 2002, Brussels.

ANNEX 4: EXTRACTS RELEVANT TO GENDER FROM THE EUROPEAN PARLIAMENT RESOLUTION ON THE COMMUNITY STRATEGY ON HEALTH AND SAFETY AT WORK 2002–06

The European Parliament examined and discussed the European Commission's Community strategy on health and safety at work 2002–06 and then adopted a resolution stating its viewpoints and proposals concerning the strategy. The Community strategy included reference to mainstreaming gender into activities on occupational safety and health and the following extracts from the European Parliament's resolution are relevant to the gender dimension in the Commission document.

'[…] whereas the gender and demographic dimensions are evoked in the Commission's analysis but are almost entirely absent from the actions proposed; for example, proposals for action should take account of the fact that many sectors of employment dominated by women are low paid and precarious; and that where men are more prone to serious industrial accidents, women tend to suffer work-related diseases and mental disorders,

3. Calls on the Commission to integrate a gender dimension throughout the strategy, paying particular attention to the following:

(i) the double workload faced by many people, predominantly women, of paid employment and socially productive but unpaid work in connection with personal responsibilities, which may have health and safety implications;

(ii) the fact that, in the context of the gender segmentation that characterises too many European labour markets, women receive less vocational training and education than their male colleagues;

(iii) the situation of specific groups such as homeworkers (a high percentage of whom are women) who manufacture a huge range of goods in their homes and women in agriculture and family-run SMEs, many of whom are legal partners in the business and carry out potentially dangerous work but have little access to training, information or social security;

(iv) how health, safety and reproductive health aspects of employment can be a hindrance to work entry by women in certain sectors which in turn perpetuates the gender division of labour and to propose measures to ensure that these obstacles are removed.

Legislative issues

4. Calls for the extension of the scope of the framework directive, Council Directive 89/391/EEC of 12 June 1989 on the introduction of measures to encourage improvements in the safety and health of workers at work to excluded groups of workers such as the military, the self-employed, domestic workers and homeworkers; the exclusion of the last two groups constitutes indirect discrimination against women and to end it would give

concrete effect to the Commission's emphasis on gender;

5. Supports in its resolution of 20 September 2001 on harassment at the workplace, the "global approach to well-being at work" including all kinds of risks such as stress, harassment, bullying and violence; this echoes the judgment of the Court of Justice as to the wide interpretation of "working environment" (Article 137(1) TEC); however, the Commission should recognise gender as important in this context and be more concrete by, for example, proposing legislation on workplace bullying; the global approach should also include a commitment to extend strategic action to address or combat future risks posed by new processes or changes to existing ones;

6. On musculoskeletal problems, which are suffered by over half the people affected by work-related complaints in the EU, particularly older workers; urges the Commission to propose a comprehensive directive on workplace ergonomics, to review and improve the implementation of Council Directive 90/270/EEC of 29 May 1990 on the minimum safety and health requirements for work with display screen equipment and to propose amendments to Council Directive 90/269/EEC of 29 May 1990 on the minimum health and safety requirements for the manual handling of loads where there is a risk particularly of back injury to workers, especially regarding repetitive handling of small loads; furthermore, gender considerations should be mainstreamed into policies and practices of risk assessment and workplace design;

[…]

10. Urges the Commission to evaluate all existing health and safety legislation with a view to establishing whether gender-specific needs have been taken into account, even in the sectors where women are under-represented;

[…]

12. Believes that legislative simplification must not become a cover for deregulation of health and safety provisions; instead, the consistency of existing provisions should be evaluated; the role of the stakeholders (especially public authorities and employee representatives) should be spelled out more clearly, ILO conventions should be better taken account of, the tasks of health surveillance should in general be better clarified than the current fragmented and inconsistent way across different directives and a full gender-mainstreaming perspective should be implemented as a horizontal issue;

[…]

30. Welcomes the Commission's new emphasis on "mainstreaming health and safety" into other Community policies and asks the Commission to include in its action plan on the strategy a clear plan for mainstreaming occupational safety and health into all relevant legislative and non-legislative areas both within its Employment and Social Affairs DG and in its other DGs; […]'.

Sources: European Parliament texts adopted at the sitting of Wednesday 23 October 2002 (P5_TA-PROV(2002)10-23, text P5_TA-PROV (2002)0499), 'Adapting to change in work and society' (A5-0310/2002 — Rapporteur: Stephen Hughes).

European Parliament resolution on the Commission communication 'Adapting to change in work and society: a new Community strategy on health and safety at work 2002–06' (COM(2002) 118 — C5-0261/2002 — 2002/ 2124(COS)).

European Parliament web site (http:// www3.europarl.eu.int/omk/omnsapir.so/ calendar?APP=PDF TYPE=PV2 FILE= p0021023EN.pdf LANGUE=EN).

ANNEX 5: STATISTICAL DATA RELATED TO GENDER AND EMPLOYMENT IN THE EU

Figure 1: Women's employment in % of total employment, 1990 and 2000 (¹)

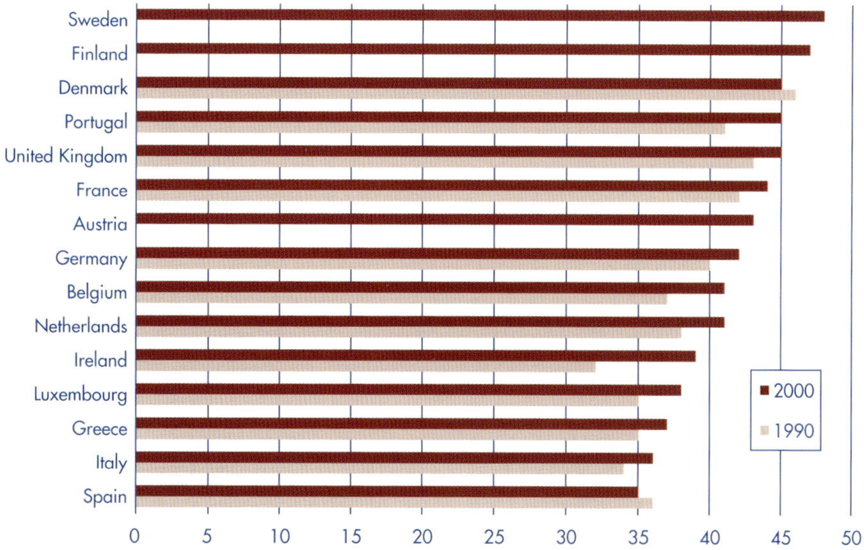
Sweden
Finland
Denmark
Portugal
United Kingdom
France
Austria
Germany
Belgium
Netherlands
Ireland
Luxembourg
Greece
Italy
Spain
0
5
10
15
20
25
30
35
40
45
50
2000
1990

(¹) In 1990, Sweden, Finland and Austria were not yet Member States and information concerning them is available for the year 2000 only.
Source: Third European survey on working conditions.

Figure 2: Employment rate of women, aged 25–49, in 1995 and 2000 in the EU (%)

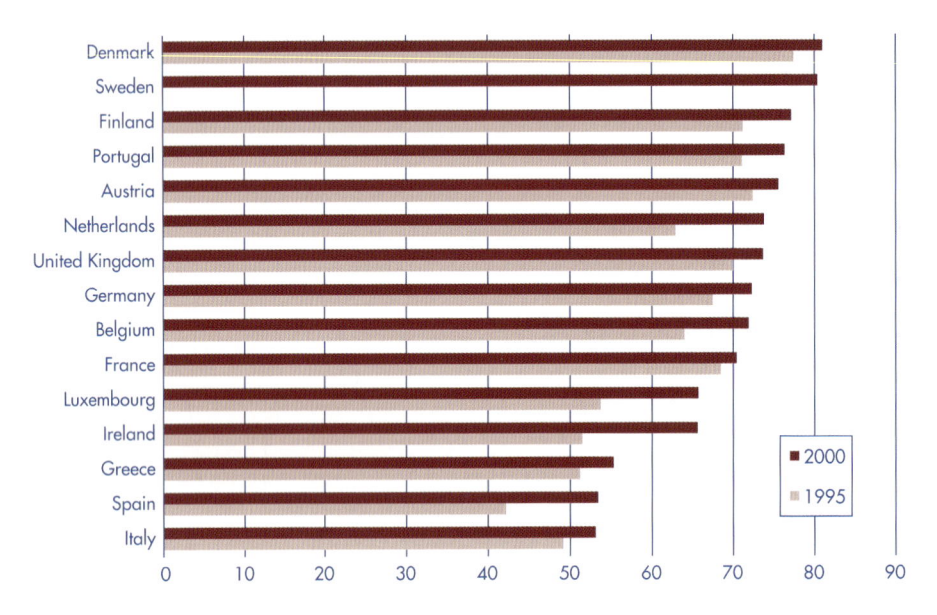

Source: Eurostat yearbook 2002.

Figure 3: Employment rate of women, aged 50–64, in 1995 and 2000 in the EU (%)

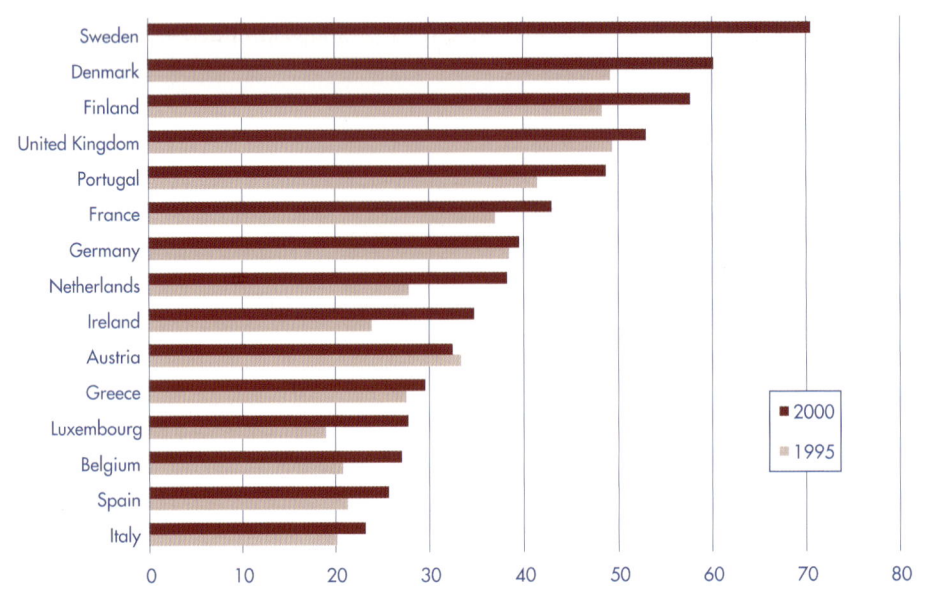

Source: Eurostat yearbook 2002.

Figure 4: Part-time employment in % of total employment, 2000

Countries (top to bottom): Netherlands, United Kingdom, Belgium, Germany, Sweden, Denmark, Austria, France, Ireland, Luxembourg, Finland, Spain, Italy, Portugal, Greece

Legend: Women, Men

Source: Key employment indicators 2000/European Commission gender equality web site: 'Statistics on gender issues — part-time employment'.

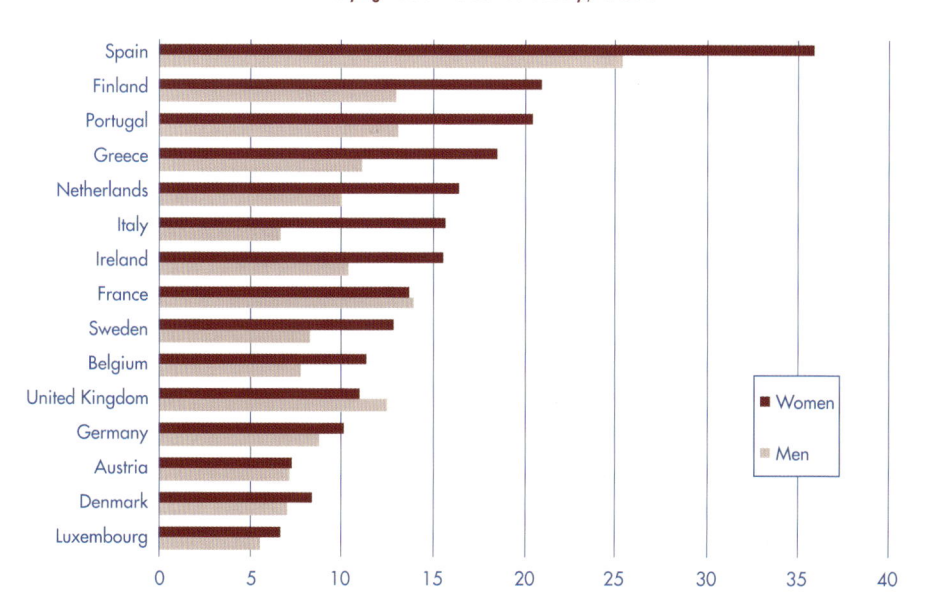

Figure 5: Percentage of employees with a non-permanent contract by gender and country, 2000

Countries (top to bottom): Spain, Finland, Portugal, Greece, Netherlands, Italy, Ireland, France, Sweden, Belgium, United Kingdom, Germany, Austria, Denmark, Luxembourg

Legend: Women, Men

Source: Third European survey on working conditions.

Gender issues in safety and health at work — A review

Figure 6: Employment status in the EU by age and gender: percentages of employees with different types of employment contracts, 2000

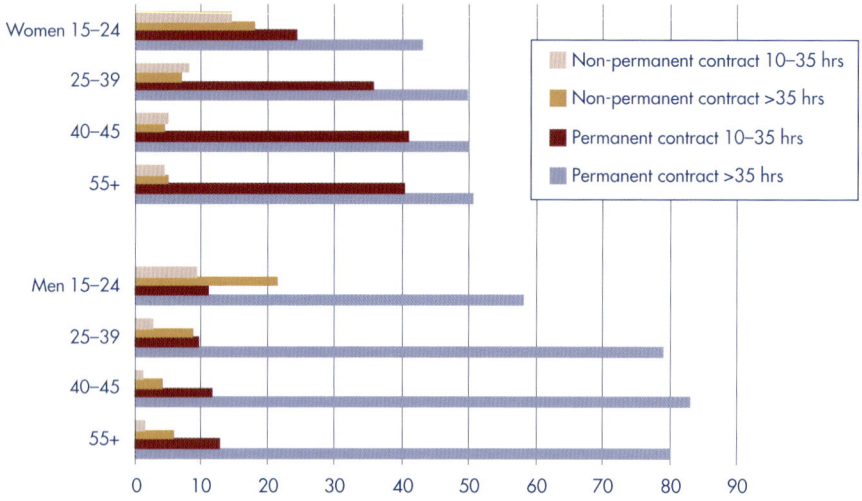

Legend:
- Non-permanent contract 10–35 hrs
- Non-permanent contract >35 hrs
- Permanent contract 10–35 hrs
- Permanent contract >35 hrs

Categories: Women 15–24, 25–39, 40–45, 55+; Men 15–24, 25–39, 40–45, 55+

Source: Third European survey on working conditions.

Figure 7: Women's employment in % of total employment by sector; services and industry, 2000

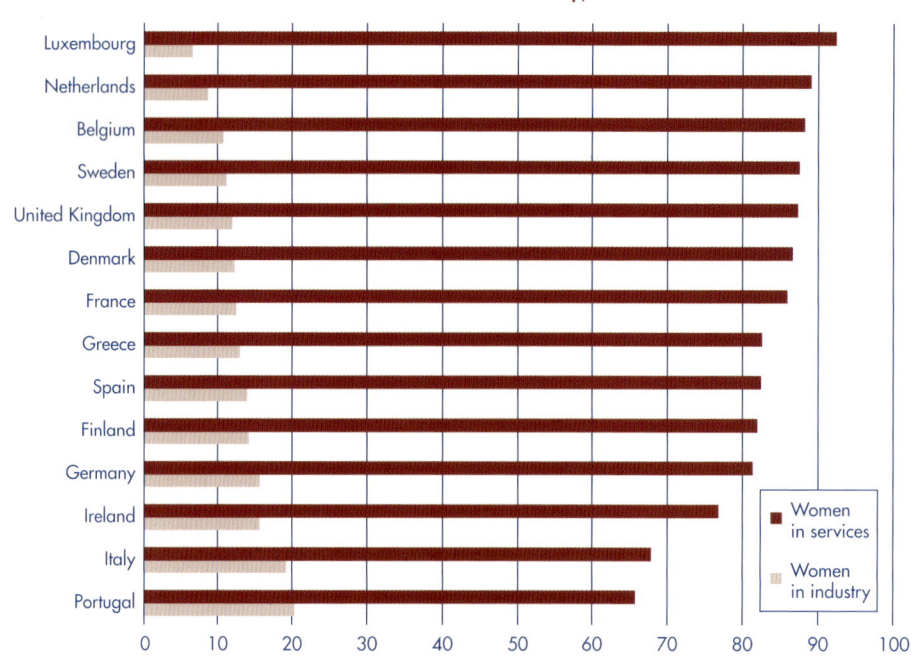

Countries (top to bottom): Luxembourg, Netherlands, Belgium, Sweden, United Kingdom, Denmark, France, Greece, Spain, Finland, Germany, Ireland, Italy, Portugal

Legend:
- Women in services
- Women in industry

Source: Employment in Europe/European Commission gender equality web site: 'Statistics on gender issues — gender segregation in the labour market'.

188

Table 1: Gender segregation by industrial sector in the EU

Percentage of the jobs in each sector that are occupied by men and women in full-time and part-time jobs

NACE sectors (¹)	Men FT	Men PT	Men All	Women FT	Women PT	Women All	Total
Construction	86	5	91	6	3	9	100
Extraction	82	2	84	16	..	16	100
Utilities	78	6	84	13	3	16	100
Transport, communications	68	7	75	16	9	25	100
Manufacturing	68	5	73	20	7	27	100
Agriculture	55	10	66	24	10	34	100
Financial services	51	7	58	27	15	42	100
Public administration	51	5	56	30	14	44	100
Sales, hotels, catering	41	6	47	30	23	53	100
Other community services	36	8	44	29	27	56	100
Health education	19	6	25	40	35	75	100
Private households and extra-territorial	2	3	5	35	60	95	100
All employment	50	6	56	26	18	44	100

Key: FT = Full-time   PT = Part-time (under 35 hours per week)   All = FT + PT
(¹) Sectors are ranked by the degree of male-dominated segregation. '..' indicates less than 0.5 %.
Source: Third European survey on working conditions.

Table 2: Women's presence in each occupational group by country, 2000

Women's percentage share of each ISCO major occupational group and of all employment, by country

	Legislators, officials and managers	Professionals	Technicians and associate professionals	Clerks	Service and shop workers	Skilled agricultural and fishery workers	Craft and related trade workers	Plant and machine operators and assemblers	Elementary occupations	All
Finland	30	40	72	87	66	44	14	15	65	49
Sweden	32	57	54	61	74	38	8	16	71	48
Netherlands	41	55	49	67	72	5	6	18	59	48
United Kingdom	37	58	37	75	75	–	7	13	45	47
Denmark	35	50	68	64	65	17	9	21	35	46
France	46	45	46	79	72	29	12	20	56	46
Portugal	42	52	30	66	58	36	23	61	71	46
Austria	38	37	57	68	78	50	11	23	51	46
Germany	25	38	58	71	74	37	11	17	51	45
Belgium	38	56	48	60	63	20	12	14	40	43
Ireland	35	51	44	73	58	20	7	38	26	43
Greece	45	54	26	61	51	46	16	3	53	41
Spain	33	49	62	68	39	25	19	11	49	39
Italy	39	54	34	51	60	25	16	14	45	38

NB: The countries have been ranked according to the female share of all employment. The armed forces, Luxembourg and agricultural work in the UK are not shown due to sample size limitations. The data are based on the original national samples, before adjusting for relative country size. The highlighted cells indicate where women are more than 5 percentage points under-represented relative to their share of all employment in the country in question. The cells printed in red are where women are more than 5 percentage points over-represented relative to their share of employment in the country in question.
Source: Third European survey on working conditions.

Table 3: Occupational segregation of women's and men's employment in the EU

Percentage of the jobs in each occupation that are filled by men and women in full-time and part-time jobs

ISCO occupational groups	Men FT	Men PT	Men All	Women FT	Women PT	Women All	Total
1. Legislators, senior officials and managers	58	5	63	30	7	37	100
11 Legislators, senior officials	71	1	72	23	6	28	100
12 Corporate managers	66	5	71	23	6	29	100
13 Managers of small enterprises	53	5	58	35	7	42	100
2. Professionals	41	9	50	27	23	50	100
21 Physical, mathematical engineering	75	6	81	16	3	19	100
22 Life science, health	30	3	33	41	26	67	100
23 Teaching	21	14	35	26	39	65	100
24 Other professionals	57	7	64	25	11	36	100
3. Technicians and associate professionals	46	6	52	31	17	48	100
31 Physical engineering associates	75	3	78	17	5	22	100
32 Life science and health associates	17	3	20	50	30	80	100
33 Teaching associates	21	10	31	30	38	69	100
34 Other associate professionals	52	8	60	26	14	40	100
4. Clerks	28	3	31	45	24	69	100
41 Office clerks	29	3	32	46	22	68	100
42 Customer services clerks	26	2	28	40	32	72	100
5. Service and shop workers	28	5	33	33	34	67	100
51 Personal protective services	30	5	35	33	32	65	100
52 Models, sales demonstrators	25	6	31	33	36	69	100
6. Skilled agricultural and fishery workers	59	10	69	22	8	31	100
7. Craft and related trades workers	83	4	87	9	3	13	100
71 Extraction, building trades	92	6	98	2	..	2	100
72 Metal, machinery-related trades	91	5	96	4	..	4	100
73 Precision, handicraft, printing	75	4	79	15	6	21	100
74 Food, textiles, wood-related	63	3	66	24	10	34	100
8. Plant and machine operators and assemblers	77	6	83	14	3	17	100
81 Stationary plant-related operators	82	3	85	15	..	15	100
82 Machine operators, assemblers	65	4	69	25	6	31	100
83 Drivers, mobile plant operators	85	10	95	3	2	5	100
9. Elementary occupations	41	9	50	22	28	50	100
91 Cleaning, domestic services, refuse, street vendors	32	9	41	22	37	59	100
92 Agricultural, fishery-related labourers	27	15	42	44	14	58	100
93 Other labourers	67	8	76	16	8	24	100
0. Armed forces	90	2	92	7	1	8	100
All employment	50	6	56	26	18	44	100

Key: FT = Full-time   PT = Part-time (under 35 hours per week)   All = FT + PT
NB: The ISCO classification is presented for the nine main occupational groups (one-digit classification) and the second-level sub-category that exists within these groups (two-digit classification). '..' indicates less than 0.5%.
Source: Third European survey on working conditions.

ANNEX 6: DATA FROM THE THIRD EUROPEAN SURVEY ON WORKING CONDITIONS

Source: These tables and figures, which present data from the third European survey on working conditions broken down by sex, are taken from Fagan and Burchell (2002).

Key to tables:
FT = Full-time; PT = Part-time (under 35 hours per week); All = FT + PT

Table 1: Intimidation and discrimination at the workplace

Percentage of respondents to the third European survey on working conditions who report that over the 12 months	Men FT	Men PT	Men All	Women FT	Women PT	Women All	All
They have personal experience of violence or intimidation	9	8	9	13	12	12	11
They know that this has been experienced by other people	15	14	15	19	19	19	17
They have personal experience of discrimination	5	6	5	8	7	8	6
They know that this has been experienced by other people	10	9	10	11	9	11	10
All	39	37	39	51	47	50	44

NB: The questions asked were 'Over the past 12 months, have you, or have you not, been subjected at work to […] by either colleagues or other people at your workplace? And 'In the establishment where you work, are you aware of the existence of […] ?' followed by a list of issues. The measure of discrimination includes sex, age, nationality, ethnic background, disability and sexual orientation.
Source: Third European survey on working conditions.

Table 2: Exposure to material and physical environmental hazards by gender

Percentage of respondents to the third European survey on working conditions who report exposure to these working conditions for at least half of their time at work	Men FT	Men PT	Men All	Women FT	Women PT	Women All	All
Loud noise	27	18	26	12	10	11	20
Vibrations from hand tools, machinery, etc.	24	14	23	9	5	7	16
Breathing vapours, fumes, dust or dangerous substances	21	15	20	11	7	9	15
High temperatures	17	15	17	11	11	11	14
Low temperatures	17	16	16	7	6	7	12
Handling dangerous products or substances	12	8	12	7	6	6	9
Radiation	4	3	4	3	1	2	3
Summary scale of ambient exposure							
High score	45	31	44	23	18	21	34
Mid-score	30	36	30	35	35	35	32
No exposure	25	33	26	42	47	44	34
Total	100	100	100	100	100	100	100

NB: The 'ambient exposure' scale is based on a summation of the seven items listed in the table, where the degree of exposure to each item was measured on a seven-point scale ranging from 'all of the time' to 'never'.
Source: Third European survey on working conditions.

Table 3: Ergonomic conditions by gender

Percentage of respondents to the third European survey on working conditions who report exposure to these working conditions for at least half of their time at work	Men FT	Men PT	Men All	Women FT	Women PT	Women All	All
Repetitive hand or arm movements	48	43	47	48	44	46	47
Painful or tiring positions	33	30	33	35	29	33	33
Carrying or moving heavy loads	28	24	28	18	16	17	23
Summary scale of hazardous ergonomic conditions							
High score	27	24	27	25	19	22	30
Mid-score	46	40	45	44	47	45	45
No exposure	27	36	28	31	34	32	25
Total	100	100	100	100	100	100	100

NB: The 'hazardous ergonomic conditions' scale is based on a summation of the three items listed in the table, where the degree of exposure to each item was measured on a seven-point scale ranging from 'all of the time' to 'never'.
Source: Third European survey on working conditions.

Table 4: Exposure to material, physical and ergonomic hazards by gender and occupational status

Percentage of respondents to the third European survey on working conditions who report a high rate of exposure to material, physical and ergonomic hazards

Occupational status group	Material and physical hazards Men	Material and physical hazards Women	Ergonomic hazards Men	Ergonomic hazards Women	Repetitive tasks of less than 10 minutes Men	Repetitive tasks of less than 10 minutes Women
White-collar managerial jobs	23	17	12	16	37	42
White-collar professional jobs	19	20	8	15	33	36
White-collar clerical and service jobs	20	13	17	19	46	47
Blue-collar craft and related manual jobs	72	49	41	48	53	68
Blue-collar operating and labouring manual jobs	60	36	43	40	55	60
All	44	21	27	22	46	48

Source: Third European survey on working conditions.

Table 5: Selective indicators of skill demands in men's and women's jobs by occupational status (¹)

Occupational status group	Percentage of employed men whose job involves: A	B	C	D	Percentage gender gap difference (women–men) (²) A	B	C	D
White-collar managerial jobs	76	46	39	73	–6	–13	–3	–13
White-collar professional jobs	81	53	33	31	0	–1	–8	–14
White-collar clerical and service jobs	64	32	36	21	–7	–8	0	–11
Blue-collar craft and related manual jobs	64	36	37	23	–17	–49	–5	–1
Blue-collar operating and labouring manual jobs	46	19	35	8	–23	–11	–8	–3
All	65	37	36	26	–4	–7	–7	–10

Key: A = Problem solving and learning
B = Only complex tasks
C = Team work and task rotation
D = Extensive planning responsibilities

(¹) These indicators were derived from questions relating to problem solving; monotonous and/or complex tasks; planning of production, staffing and/or working time; task rotation and/or team work.
(²) The 'gender gap' difference is the percentage point difference obtained by subtracting the male score from the female score. A negative sign indicates that fewer women are employed in this occupational status group report.

Source: Third European survey on working conditions.

Table 6: The degree of autonomy in men's and women's jobs by occupational status

Work autonomy	Percentage with a high degree of task autonomy					
	Men			Women		
	Low	Some	Higher	Low	Some	Higher
White-collar managerial jobs	7	20	73	8	25	68
White-collar professional jobs	10	37	53	14	50	36
White-collar clerical and service jobs	24	38	39	30	37	33
Blue-collar craft and related manual jobs	28	33	40	25	28	47
Blue-collar operating and labouring manual jobs	45	31	23	40	34	27
All	25	33	42	25	39	36

Working-time autonomy	Percentage with a high degree of time autonomy					
	Men			Women		
	No	Some	Both	No	Some	Both
White-collar managerial jobs	9	24	67	9	25	66
White-collar professional jobs	25	28	47	41	30	29
White-collar clerical and service jobs	32	35	33	34	37	29
Blue-collar craft and related manual jobs	34	33	33	34	26	41
Blue-collar operating and labouring manual jobs	44	37	20	48	31	21
All	31	32	37	36	33	31

NB: The 'work autonomy' scale includes work methods, speed of work, task order and breaks. Low autonomy refers to autonomy on one or none of the four items listed; higher refers to autonomy on all four items.
Source: Third European survey on working conditions.

Table 7: Perceptions of the health impacts on employment and absenteeism rates

Percentage of respondents to the third European survey on working conditions who report the following	Men			Women			All
	FT	PT	All	FT	PT	All	
Job affects their health in some way	62	55	61	61	53	58	60
Work improves health	1	1	1	1	2	2	1
Health or safety is at risk because of their job	32	25	31	24	19	22	27
Absenteeism in the last 12 months							
For at least one day due to an accident at work	10	6	9	6	5	6	8
If absent: the average number of days absent	21	15	22	21	23	21	21
For at least one day due to health problems caused by work	10	7	10	12	9	10	10
If absent: the average number of days absent	18	19	18	17	25	20	19
For at least one day due to other health problems	34	29	33	40	34	36	35
If absent: the average number of days absent	11	14	12	13	13	13	12

Source: Third European survey on working conditions.

Table 8: Men's and women's assessments of the health impact of their jobs

Percentage of respondents to the third European survey on working conditions who report that they have the following health problems due to their working conditions	Men FT	Men PT	Men All	Women FT	Women PT	Women All	All
Hearing problems	11	6	10	3	2	3	7
Vision problems	9	8	9	10	5	8	8
Allergy-related problems (allergies, skin problems, respiratory difficulties)	10	8	10	9	7	8	9
Muscle-related problems (backache and other muscular pains)	30	24	29	33	26	30	30
Stress-related problems (headaches, stomach aches, heart disease, anxiety, irritability, sleeping problems, 'stress')							
One symptom	20	17	19	18	16	16	19
At least two symptoms	25	23	25	29	19	27	25
Total with some stress symptoms	45	40	44	47	35	43	44

Source: Third European survey on working conditions.

Table 9: The domestic responsibilities of employed men and women in the EU

Household situation	contributes most to household income (1)	is mainly responsible for housework and shopping (2)	looks after children or elderly relatives every day (3)
Men			
Married/cohabiting parent	91	11	57
Married/cohabiting, no child at home	89	12	11
Sole adult, no child at home	53	43	4
All employed men	77	20	25
Women			
Married/cohabiting parent	19	93	88
+ employed full-time	24	92	86
+ employed part-time	13	95	89
Married/cohabiting, no child at home	22	91	23
Sole adult, no child at home	56	59	11
All employed women	31	83	45

(1) 'I am the person in my household who contributes most to the household income'.
(2) 'I am the person in my household who is mainly responsible for ordinary shopping and looking after the home'.
(3) 'I spend at least one hour a day looking after children or elderly relatives'.
NB: Dependent children are defined as aged under 15 years and living in the same household. Lone parents are not shown separately but are included in the overall figures.
Source: Third European survey on working conditions.

Table 10: Other activities beyond employment and domestic responsibilities undertaken by employed men and women in the EU

Household situation	Percentage of respondents to the third European survey on working conditions who regularly participate in:		
	leisure, sport or cultural activities during the week (¹)	education or training courses outside of work (²)	civic activities during the year (³)
Men			
Married/cohabiting parent	66	9	34
Married/cohabiting, no child at home	64	8	32
Sole adult, no child at home	79	12	25
All employed men	69	10	31
Women			
Married/cohabiting parent	57	10	29
+ employed full-time	*53*	*10*	*28*
+ employed part-time	*62*	*10*	*31*
Married/cohabiting, no child at home	61	11	31
Sole adult, no child at home	71	22	29
All employed women	62	13	30

(¹) Leisure, sport or cultural activity during the week (note that it was left to the respondents to define what they considered to be 'leisure' or 'cultural' activities, and leisure may include relaxing or watching TV as well as more structured activities).
(²) Education or training courses at least once a month.
(³) Involved in voluntary, charitable, political or trade union activity during the year.
NB: Dependent children are defined as aged under 15 years and living in the same household. Lone parents are not shown separately but are included in the overall figures.
Source: Third European survey on working conditions.

ANNEX 7: DATA FROM EUROSTAT ON THE HEALTH AND SAFETY OF MEN AND WOMEN AT WORK

The following tables are taken from *Statistics in Focus*, Theme 3: Population and social conditions, 4/2002 (Dupré, 2002), Eurostat, European Commission.

The European statistics on accidents at work are based on records collected by Member States for work accidents resulting in four or more days' absence from work. The work-related illhealth data come from responses to the 1999 European labour force survey.

The incidence rate is defined as the number of accidents at work occurring during the year per 100 000 persons in employment. The prevalence rate is the number of work-related health complaints suffered over the past 12 months per 100 000 persons in employment. Here, the incidence rate is measured in terms of full-time equivalent employment (see below) in order to allow for differences in hours worked both between men and women and between jobs in different sectors of activity. Full-time equivalent (FTE) employment of men and women is estimated by adjusting the number employed in each sector or occupation in the different Member States by the average hours they usually work per week relative to the average hours worked by men and women employed full-time in the EU as a whole. This differs from the usual procedure of defining FTE employment in terms of the usual full-time hours worked in each country separately because the aim is to adjust for differences in working time between Member States as well as between men and women, types of activity and occupations. Equivalent information is not available from all Member States. For further explanation, see Dupré (2002).

Gender issues in safety and health at work — A review

Figure 1: Incidence rate of accidents at work suffered by men relative to women, 1998

Accidents with more than three days' absence per 100 000 employed, ratio of men to women
NL: no data by sex available

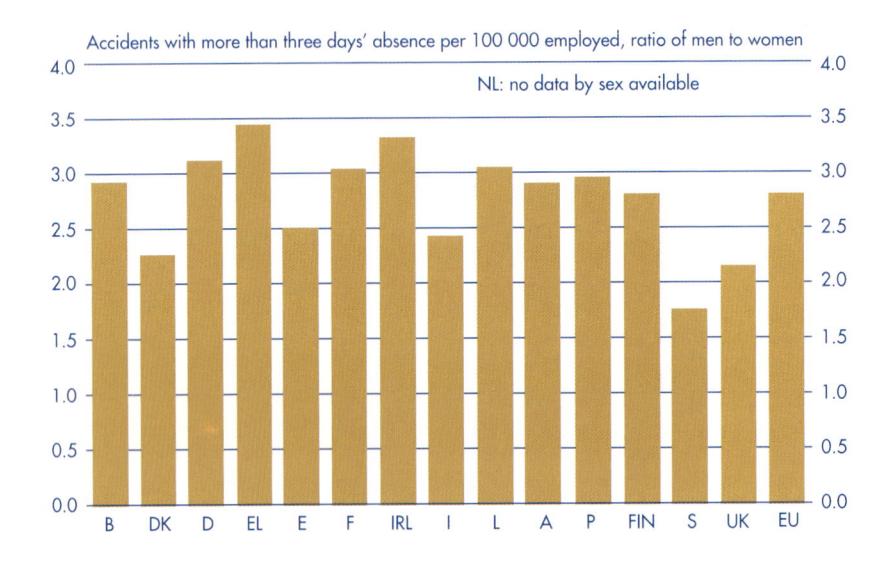

Figure 2: Incidence rate of accidents at work suffered by men relative to women on an FTE basis, standardised for sectoral structure, 1998 (more than three days' absence)

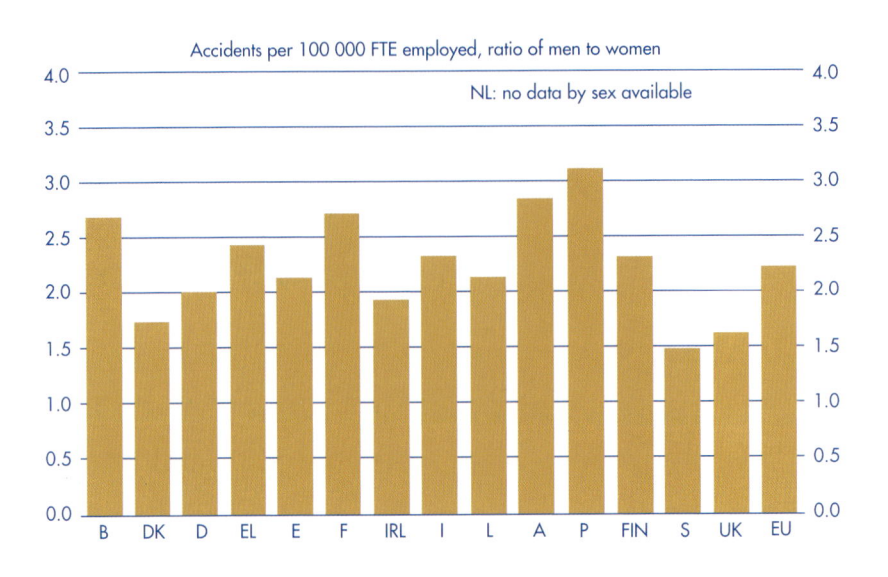

Accidents per 100 000 FTE employed, ratio of men to women
NL: no data by sex available

Figure 3: Standardised incidence rates of accidents at work to men and women in relation to FTE employment by branch in the EU, 1998 (more than three days' absence)

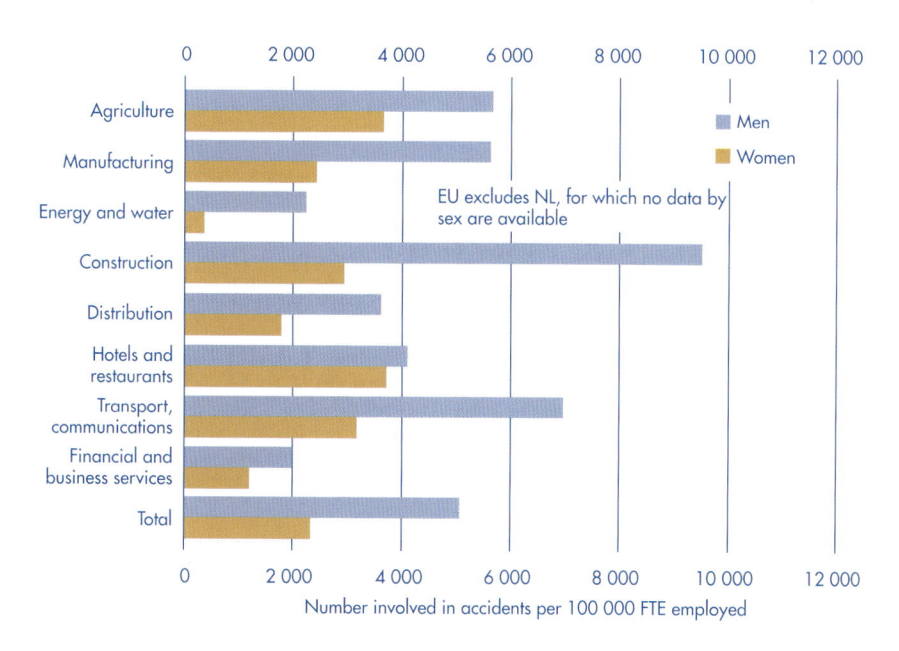

Figure 4: Change in standardised incidence rates of accidents at work to men and women, 1994–98 (more than three days' absence)

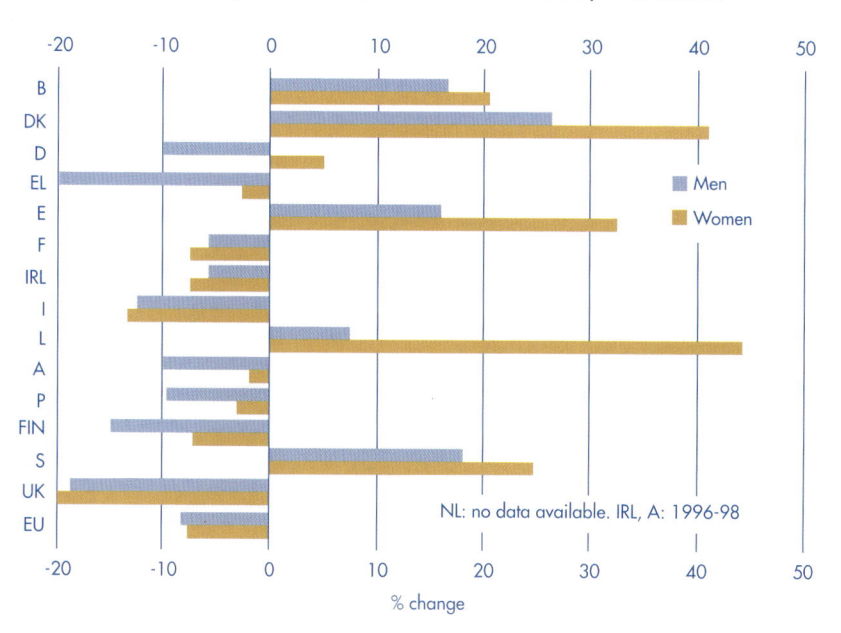

Table 1: Incidence rates of accidents at work to men relative to women by sector of activity, 1998

Number involved in accidents at work with more than three days' absence per 100 000 FTE employed, ratio of men to women

Sector	B	DK	D	EL	E	F	IRL	I	L	A	P	FIN	S	UK	EU
Agriculture	1.6	1.0	1.5	3.7	1.2	1.4	2.3	1.2	0.9	2.0	3.1	2.7	0.9	1.7	1.5
Manufacturing	2.7	1.7	2.0	2.4	2.4	2.4	1.6	3.0	1.4	2.6	3.8	2.4	1.5	1.9	2.3
Energy and water	2.5	3.8	22.4	8.9	3.2	5.1	:	3.3	:	2.1	0.7	6.4	3.2	8.3	6.7
Construction	8.8	4.4	2.4	2.5	2.6	11.1	1.3	6.4	:	7.4	2.4	3.6	3.8	4.2	3.3
Distribution	2.3	1.6	2.6	2.0	2.1	2.0	2.0	2.5	2.2	2.3	3.2	1.6	1.1	1.0	2.1
Hotels and restaurants	0.9	1.1	1.1	1.4	1.0	1.1	0.6	0.6	0.9	1.4	7.7	1.7	0.7	0.9	1.1
Transport, communications	2.9	1.1	2.2	3.6	2.7	3.0	4.9	1.9	4.0	2.7	1.1	2.2	1.4	2.0	2.2
Financial and business services	2.4	1.2	0.6	2.9	1.3	2.5	1.3	1.4	2.3	2.8	4.2	1.8	1.3	1.2	1.7
Total	2.7	1.7	2.0	2.4	2.1	2.7	1.9	2.3	2.1	2.8	3.1	2.3	1.5	1.6	2.2

NB: EU excludes NL, for which data by sex are not available.

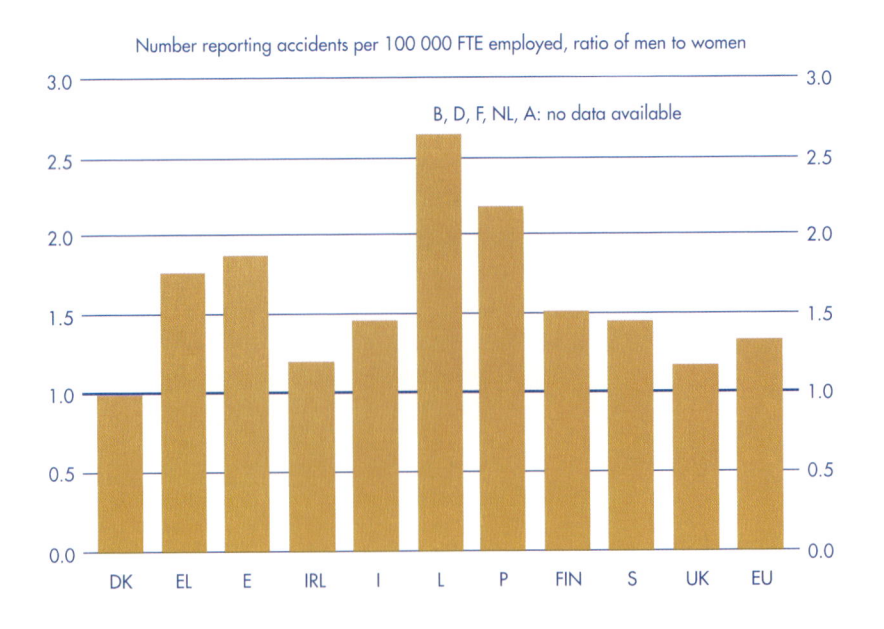

Figure 5: Incidence rates of accidents at work to men relative to women standardised for occupational structure, 1999

Number reporting accidents per 100 000 FTE employed, ratio of men to women

B, D, F, NL, A: no data available

European Agency for Safety and Health at Work

Figure 6: Prevalence rates of work-related complaints of men relative to women, standardised (occupation), 1999

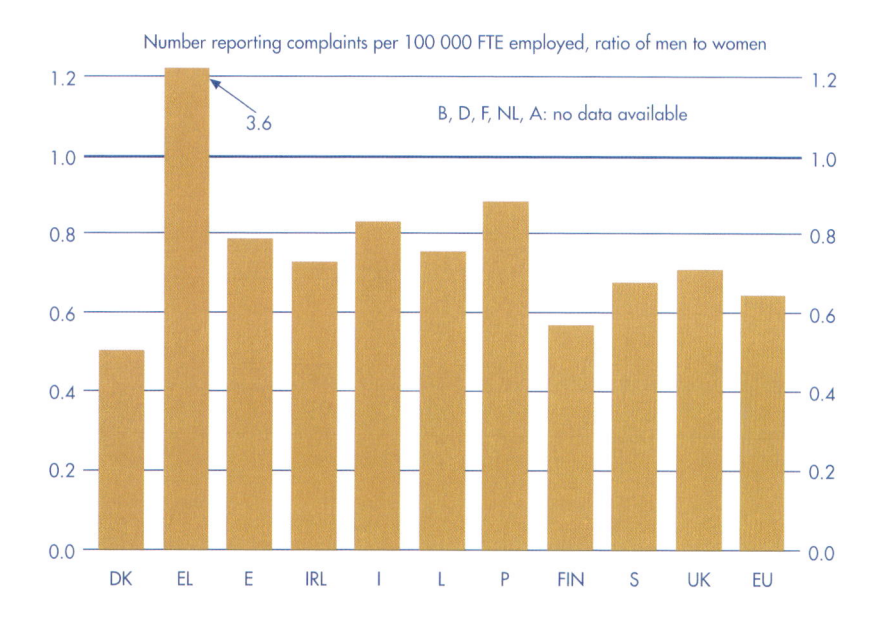

Number reporting complaints per 100 000 FTE employed, ratio of men to women

B, D, F, NL, A: no data available

EL: 3.6

Table 2: Prevalence rates of work-related complaints by occupation, 1999

	Ratio of men to women									
Main occupations (ISCO)	DK	EL	E	IRL	I	L	P	FIN	S	UK	EU
Managers, professionals, technicians	0.4	1.0	0.8	0.7	0.7	0.7	0.7	0.6	0.4	0.6	0.5
Office workers	0.9	:	1.3	1.1	0.9	0.8	0.8	0.6	1.0	0.8	0.7
Sales and service workers	0.5	:	0.7	0.6	1.0	0.8	0.5	0.4	0.4	0.6	0.5
Skilled manual workers	0.5	2.4	0.6	1.1	0.9	0.6	0.9	0.6	0.6	1.0	0.8
Elementary workers	0.5	1.4	0.9	0.5	0.7	2.1	1.3	0.5	1.1	0.8	0.7
Standardised total	0.5	3.6	0.8	0.7	0.8	0.8	0.9	0.6	0.7	0.7	0.6

NB: EU excludes B, D, F, NL and A, for which data or breakdown by occupation are not available.

Table 3: Proportion of men and women affected by different types of work-related complaints, 1999

Percentage of men/women by type of complaint reported as being most serious

Type of complaint	DK	EL	E	I	L	P	FIN	S	UK	EU
Men										
Bone, joint or muscle problem	57.3	38.8	53.0	50.3	44.3	45.6	58.6	59.7	44.1	51.4
Stress, depression or anxiety	8.4	10.7	7.3	12.6	7.3	15.2	11.2	14.2	30.5	16.5
Breathing or lung problem	4.8	17.5	12.6	10.3	12.6	11.3	11.8	5.8	3.7	8.4
Heart disease or attack, or other problems of the circulatory system	2.5	0.0	11.2	5.4	9.0	6.3	5.1	3.6	3.2	5.4
Hearing problem	1.9	4.9	2.3	8.3	2.5	4.5	4.2	3.8	2.5	4.2
Headache and/or eye strain	3.3	9.2	2.1	4.4	7.7	2.3	1.6	1.7	2.9	2.8
Skin problem	1.6	14.1	1.1	3.2	3.7	4.4	2.9	1.0	2.5	2.4
Infectious disease (virus, bacteria or other type of infection)	3.3	0.0	1.6	3.0	6.1	2.2	1.2	1.2	2.8	2.3
Other types of complaint	16.8	4.9	8.7	2.5	6.8	8.2	3.3	9.0	7.8	6.7
Total	100.0	100.0	100.0	100.0	100.0	100.0	100.0	100.0	100.0	100.0
Women										
Bone, joint or muscle problem	63.4	:	66.1	48.3	33.6	26.8	63.9	60.7	40.4	54.4
Stress, depression or anxiety	9.3	:	8.7	17.0	13.7	34.3	11.5	20.6	36.5	20.2
Breathing or lung problem	2.5	:	5.3	9.4	13.6	13.1	10.4	3.2	4.5	6.4
Headache and/or eye strain	3.9	:	2.1	6.6	8.0	7.5	2.5	1.7	4.5	3.7
Infectious disease (virus, bacteria or other type of infection)	3.1	:	1.4	5.3	9.1	0.8	1.4	1.9	3.6	2.8
Skin problem	2.3	:	1.3	3.7	1.1	4.2	3.7	2.3	1.9	2.6
Heart disease or attack, or other problems of the circulatory system	1.0	:	6.1	3.3	7.4	1.9	2.8	1.4	1.4	2.5
Hearing problem	1.4	:	0.4	2.8	0.4	0.5	0.8	0.8	0.4	1.0
Other types of complaint	13.1	:	8.7	3.6	13.0	10.9	2.9	7.4	6.8	6.4
Total	100.0	:	100.0	100.0	100.0	100.0	100.0	100.0	100.0	100.0

NB: EU excludes B, D, EL (women only), F, IRL, NL and A, for which data or breakdown by sex are not available.

ANNEX 8: OCCUPATIONAL HAZARDS AND GENDER WORLDWIDE

At the international level, the International Labour Organisation (ILO) and others have carried out activities to analyse gender issues in relation to occupational hazards.

On average, women's share of the global labour market is over 40 %. The general trend is that the number of women in paid work is growing. However, in most countries, the unemployment rate is higher for women than for men. A majority of the world's women still work in agriculture, often as unpaid family workers, but many have found new job opportunities with better pay in the industry and service sectors. Women are increasingly active as entrepreneurs and own or operate a significant proportion of small and micro-businesses in many countries. Furthermore, women have moved up the ladder to take up an increasing number of professional and managerial positions. However, they remain clustered in the lower and middle levels of firms and with few exceptions rarely rise to top positions in either the private or public sector.

Progress towards gender equality is uneven and discrimination continues to be pervasive in the world of work. Women still tend to be disproportionately concentrated in the lower skilled and lower paid jobs. They form the majority of workers carrying out work on a part-time, temporary or home-based basis.

The occupational hazards affecting women are related to the type of work they typically perform. These include lack of safety and health measures and services combined with long hours of work, particularly in the case of homework. The injuries and reproductive risks caused by exposure to harmful pesticides and heavy physical work during crop cultivation and harvesting are also common. In addition, women in all walks of life have to assume greater responsibility for the care work of family and home, although a gradually growing number of men in a range of countries are beginning to play a greater role in this regard. Women's higher stress levels, chronic fatigue and premature ageing can be attributed to the 'double day' or the effort of balancing work and family demands.

Women in agriculture, like many other rural workers, have a high incidence of injuries and diseases and are insufficiently reached by health services. Women's role in agriculture has been traditionally underestimated. Today, women produce almost half of the world's food. The average earnings of rural women engaged in plantation work are lower than those of men. Many women in agricultural labour end up doing jobs that nobody else would do, such as the mixing or application of harmful pesticides without adequate protection and information, suffering from intoxication and, in some cases, death. Heavy work during crop cultivation and harvesting can result in a high incidence of stillbirths, premature births and death of the child or the mother. Some studies have shown that the workload of traditional 'female' tasks, such as sowing out, picking out, and clearing, is a little higher than the workload of males due to the fact that the latter are assisted by mechanical means during irrigation, ridging and farming (ILO web site).

Gender discrimination also makes it harder for women to break out of the cycle of poverty, particularly for households headed by women, which are becoming increasingly common in many countries. Women's higher levels of illiteracy and lack of access to education and training in comparison with men, as well as their lack of representation in political organisations, adversely affect their occupational choice and earnings. It is important to address these issues, just as it is important to analyse and recognise the different hazards and risks experienced by men and women in the globalising world of work and local labour market. A clear policy commitment is needed to combat discrimination, i.e. to ensure that both men and women benefit from occupational safety and health programmes, legal provisions and institutions established for their implementation and enforcement (Forastieri, 2000; see also 'ILO information note' below).

Global strategies and policies on occupational safety and health should also take specific account of the additional problems women workers face particularly in the developing world, such as heavy physical work, the reconciliation of work and family, work in the context of the family, less developed working methods, traditional social roles, and poverty. General poverty and illiteracy rates among women workers in the developing world are even higher than for men (WHO, 1994a; Kane, 1999a; see also Bildt et al., 2002).

The ILO has set an agenda called 'Decent work' to respond to the challenge presented by globalisation, and to promote opportunities for women and men to obtain decent and productive work. The agenda's objective is to set a threshold for work and employment which embodies universal rights and which is consistent with a society's values and goals. The 'Decent work' initiative puts gender equality and development issues at the heart of the ILO agenda (ILO, Bureau for Gender Equality, 2000).

Women and pesticides in Ukraine

Women in Ukraine constitute 54 % of the rural population and undertake 65 % of agricultural labour. Their employment conditions are worse than those of men and they are widely involved in pesticide application, including obsolete pesticides. Work may include digging land contaminated by radionuclides and pesticide residues in order to plant potatoes, and spraying pesticides without wearing protective clothing. There is evidence that they are suffering pesticide-related diseases but the women themselves are largely ignorant of the dangers. Intensive production remains the favoured policy, attracting the major agrochemical corporations.

One initiative to mainstream gender into activities to protect the environment and to promote sustainable development, with women taking the lead, comes from the non-governmental organisation, the Centre of Sustainable Development and Ecological Research, and is entitled 'Women in agrarian education'. Activities include the promotion of sustainable agriculture, pesticide reduction as a key focus, environmental education and awareness raising, and promoting the adoption of legislation and policies for pesticide reduction strategies at local and State levels.

Source: Stefanovska and Pidlisnuk (2002).

Examples of occupational safety and health strategy and policy worldwide

In general, the need to pay attention to gender issues is being increasingly recognised and similar approaches are being advocated, particularly the mainstreaming of gender into occupational safety and health policy and activities. Examples are given below from international bodies.

ILO/WHO Joint Committee

The ILO/WHO Joint Committee recognised in 1992 that there are specific occupational health needs of workers because of age, physiological conditions, gender, communication barriers and other social aspects, among other factors. It advocates, as a priority field of action, the development of activities in which such needs be met on an individual basis with due concern for the protection of all workers' health at work, without leaving any possibility for discrimination (ILO, 1992).

ILO

The International Labour Conference resolution on equal opportunities and equal treatment for men and women in employment of 1985 (see below) includes items on the working environment, for example, that 'women and men should be protected from risks inherent in their employment and occupation in the light of advances in scientific and technological knowledge' and that attention should be paid to protecting the reproductive health of both men and women.

In 1989, the ILO organised a tripartite meeting of experts on 'special protective measures for women and equality of opportunity and treatment'. The main conclusion of the discussions was that both women and men require appropriate protection against occupational safety and health hazards and poor working conditions. Special protective measures for women are incompatible with the principle of equality of opportunity and treatment unless they arise from women's biological condition. More research is needed into the differences in the reproductive function of women and men. Other points made during the meetings included: the need for change in the basic attitudes and practices of society regarding equal opportunity and treatment; the recognition of women's additional work burdens; the need to review all protective legislation applying to women; the need to justify all protective legislation with objective and up-to-date scientific evidence; the need for reasonable accommodation, hours of work and rest periods, and washing and sanitary facilities during pregnancy, nursing and early maternity; the need for personal security for all workers (including against sexual harassment and violence); and the importance of childcare facilities for working men and women.

Examples of gender-sensitive ILO conventions

Convention 171 on Night Work (1990) shows a shift from the earlier concept of protecting all women to an understanding that protection should be granted to all workers. Women's general exclusion from night work could have a negative effect on their chances in the labour market and is increasingly seen as discriminatory unless justified by reasons linked with maternity.

Convention 176 on Home Homework (1996), which addresses a female-dominated area of work (studies show that around 90 % of homeworkers are women), sets out the following provisions: freedom of association, protection against discrimination, equal remuneration for work of equal value, statutory social security protection, occupational safety and health, minimum age and maternity protection, and access to training.

Convention 183 on Maternity Protection (2000). What is new in this recent standard is the extension of coverage to all employed women, stronger protection from dismissal during pregnancy or maternity leave and after return to work, health protection for pregnant and breastfeeding mothers and measures to ensure that maternity is not a reason for discrimination in employment.

Convention 184 on Safety and Health in Agriculture stresses the importance of protecting young and women workers, acknowledging the high number of women involved in agricultural work.

Since the passing of the 1985 resolution and the 1989 meeting of experts, the ILO adopted in the mid-1990s an in-house gender mainstreaming policy, which includes several components. One is a requirement that all ILO staff carry out gender analyses of the labour market and take differences between women and men in the labour market into account in the design and implementation of technical programmes. The ILO has also appointed gender specialists in many field offices and established a broad gender network for more effective gender mainstreaming in its work.

Another consequence of the ILO's shift to a more gender-focused approach during the past 20 years is the adoption of several 'gender-sensitive' ILO conventions, including on night work, homework, maternity protection and agriculture (see box above).

The ILO is also addressing sexual harassment at work. The most recent activities include national studies and a regional meeting in Asia, which resulted in the publication *Action against sexual harassment at work in Asia and the Pacific* (2001). In the field of preventing psychosocial problems at work, a training programme called SOLVE addresses stress and violence, including sexual harassment.

Specifically for occupational safety and health, the ILO has produced suggestions on how to integrate the gender perspective into the field of occupational safety and health (see also 'ILO information note' below). Among these suggestions are the following issues:
- special consideration to women's safety and health in all occupational safety and health policies;
- targeting occupational hazards to which women are exposed at the enterprise level;
- ergonomic considerations based on a non-discriminatory approach and individual risk assessment;
- planning for human variability, i.e. national standards providing protection against all hazards for all workers regardless of age or sex;
- research evaluating real differences between the sexes;
- data collection;
- women's participation.

Global Commission on Women's Health

The World Health Organisation (WHO), through a wide-ranging resolution concerning its work in relation to women's health, established the Global Commission on Women's Health. An issues and policy paper on women and occupational health (Kane, 1999a) was prepared for the Global Commission on Women's Health which made a number of recommendations, given below. It, too, recommends gender assessment of occupational safety and health legislation:

1. Issues of women's occupational health should be examined within the context of gender-specific analyses of occupational health.
2. Such gender-specific analyses should identify the specific occupational health risks of particular industries, occupations and tasks not only for the individual worker but for other family members.
3. Women's work in the informal sector, in agriculture and in the home has to be conceptualised and measured if the specific occupational health risks of women are to be addressed.
4. The use of methodologies such as time–use surveys and record linkage in longitudinal studies to identify and assess occupational health risks should be extended.
5. Legislation addressing women's occupational health needs should be reassessed to ensure that it neither discriminates against women nor overlooks potential occupational health risks among men.
6. International agreement about the classification of reproductive hazards (such as chemicals) and on the precautions needed

to protect both men and women against those hazards should be developed.
7. The need for greater priority in addressing the occupational health needs of both men and women requires commitment and close collaboration on the part of the various international agencies concerned, such as the WHO and ILO.
8. Interdisciplinary research with a strong social science component is essential for the understanding of gender-related issues in occupational health. The WHO and its appropriate collaborating centres should take the lead in identification and coordination of such research.

ILO Conference 1985 resolution on equality: working conditions (extract)

Working conditions and environment (extract)

- 'Measures to improve working conditions and environment for all workers should be guided by the conclusions concerning future action in the field of working conditions and environment adopted by the International Labour Conference in 1984, and in particular taking into consideration the provisions concerning hygiene, health and safety at work for women. Due attention should be paid:
  — in particular to those sectors and occupations employing large numbers of women;
  — to the need to ensure proper application of relevant measures to all enterprises covered;
  — to the desirability of extending the scope of such measures so that working conditions in sectors or enterprises hitherto excluded, such as export processing and free trade zones, may be appropriately regulated;
  — to the need for national legislation to ensure that part-time, temporary, seasonal and casual workers as well as home-based workers, contractual workers and domestic workers suffer no discrimination as regards terms and conditions of employment and that further segregation of the labour market does not result;

- As regards protective legislation:
  — women and men should be protected from risks inherent in their employment and occupation in the light of advances in scientific and technological knowledge;
  — measures should be taken to review all protective legislation applying to women in the light of up-to-date scientific knowledge and technological changes and to revise, supplement, extend, retain, or repeal such legislation according to national circumstances, these measures being aimed at the improvement of quality of life and at promoting equality in employment between men and women;
  — measures should be taken to extend special protection to women and men for types of work proved to be harmful for them, particularly from the standpoint of their social function of reproduction, and such measures should be reviewed and brought up to date periodically in the light of advances in scientific and technological knowledge;
  — studies and research should be undertaken into processes which might have a harmful effect on women and men from the standpoint of their social function of reproduction, and appropriate measures, based on that research, should be taken to provide such protection as may be necessary.

- Sexual harassment at the workplace is detrimental to employees' working conditions and to employment and promotion prospects. Policies for the advancement of equality should therefore include measures to combat and prevent sexual harassment.'

Taken from the resolution on equal opportunities and equal treatment for men and women in employment, adopted by the International Labour Conference at its 71st session in June 1985.

ILO information note: How to integrate the gender perspective into the field of occupational safety and health

1. An occupational safety and health policy

If health-promotion policies are to be effective for both women and men, they must be based on more accurate information about the relationship between health and gender roles. Women workers are particularly disadvantaged by out-of-date workforce structures, workplace arrangements and attitudes. Health-promotion policies for working women need to take into account all their three roles: as housewives, as mothers and as workers. The effects on health of each role have to be looked at separately and the potential conflicts and contradictions between them need to be examined. A broad strategy for the improvement of women workers' safety and health has to be built up within a national policy on occupational safety and health, particularly in those areas where many women are concentrated. A coherent framework should be developed to ensure a coordinated national approach.

The concentration of women workers in particular occupations leads to a specific pattern of injury and disease. General measures directed to all workers do not necessarily achieve the desired benefits for women workers. The effects of gender on health need to be more carefully explored to develop a better understanding of the relationship between women's health and the social and economic roles of women. The findings need to be incorporated into policy-making.

The policy should include the specific protection of women workers' safety and health as a goal. Guidance should be provided to enable employers, trade unions and national authorities to identify problems, make the appropriate links with general safety and health activities for all workers and develop specific programmes to ensure that the needs of women workers are taken into account in occupational and industrial restructuring processes at the national level, particularly in the areas of legislation, information and training, workers' participation and applied research.

2. Targeting at the enterprise level

Industries and occupations which have an impact on the health of women workers should be key targets for change. Therefore, specific preventive programmes should be implemented. At the level of the enterprise, measures should be taken to control occupational hazards to which women workers are exposed. For the effective prevention and control of these hazards, special action programmes should be developed for work-related hazards within each occupation, including psychosocial and organisational factors, taking into consideration the physical, mental and social well-being of women workers. Revision of work practices and job redesign to eliminate or minimise hazards, job classifications, upgrading skills, and provision of new career paths in occupations where women are predominant should receive priority.

3. Targeting at the individual level

There is a need to focus on women's occupational safety and health, protecting their well-being through occupational health services. Preventive programmes need to be established to maintain a safe and healthy working environment. Work should be adapted to the capabilities of women workers in the light of their state of physical and mental health, for example by reducing women's workload, by promoting appropriate technology, by reassignment to another job according to the worker needs, and by providing rehabilitation when necessary.

Special measures for performance of physical tasks during pregnancy and childbearing are still necessary; in particular, the protection of pregnant women for whom night work, arduous work and exposure to radiation might present unacceptable health risks. However, the approach should be the equal protection against hazards in the workplace to all workers, encouraging more equal sharing of the workload between women and men in all spheres, including childcare, domestic chores and work outside the home.

4. Ergonomic considerations

The concept of maximum weight to be manually handled by women and the design of personal protective equipment need to be revised in the context of current technical knowledge and socio-medical trends. Intra-sex variations need to be taken into account.

National standards for manual handling should move away from regulating weight limits which differ between women and men workers and adopt a non-discriminatory approach based on individual risk assessment and control. Australia, Canada, and the United States are some of the countries which have introduced this criterion in their own standards.

With the worldwide massive migration, it is becoming more and more evident that anthropometric standards need to be based on human variability more than on 'model' populations, as different racial and ethnical morphological characteristics can be found among the workers of any single country.

5. Planning for human variability

Broad generalisations about women's physical capacities should be avoided and the vulnerability and needs of male workers should also be realistically taken into account. The individual capability of workers independently of age and sex should be the parameter for the performance and demands to be placed on the individual worker. Therefore, standards at national level should be adopted to provide adequate protection (against any hazard) for the most susceptible or vulnerable workers of any age or sex.

Single standards of exposure to physical, chemical or biological agents would avoid discrimination and guarantee the protection of all workers' health. Special legal protection for women should not be invalidated but should be extended to male workers where appropriate; for example, in the case of radiation protection and reproductive health.

6. Research

Existing epidemiological research must be critically assessed to find any systematic bias in the way investigation is done when studying women's health and illness patterns, to avoid assumptions based on traditional cultural values (for example, associate certain cervical cancers with certain female occupations). Evaluating real differences between the sexes and avoiding erroneous judgments about women's lives are the only way to succeed in producing knowledge which is beneficial to women's health.

7. Data collection

Similarly, national statistics on occupational accidents and diseases of women are deficient, and knowledge about women's health is still insufficient. Most countries continue to emphasise official statistics on maternal mortality, which is still a very important indicator of the general health of women in developing countries. However, many women work only part-time or are employed as homeworkers so as to be able to deal with their family responsibilities while contributing at the same time to the economy of the family. This situation excludes them from statistics on injury compensation or on absence from work because of illness. Domestic and household work is also unlikely to be recorded in any statistics. Women's occupations are often missing from medical reports or death certificates as in the case of many workers.

The development of national statistics on occupational accidents and diseases by gender would contribute: to determining priorities for action through preventive programmes; to the development of a national information strategy to collect and disseminate information on the occupational health and safety of women workers; and to the development of national standards, national codes of practice and other guidelines on specific hazards faced by women workers.

8. Women's participation

Women should be better represented and more directly involved in the decision-making process concerning the protection of their health. Women's views as users, care givers and workers, and their own experiences, knowledge and skills should be reflected in formulating and implementing health-promotion strategies. They should have greater participation in the improvement of their working conditions, particularly through programme development, provision of occupational health services, and access to more and better information, training and health education. The support of women workers to organise themselves and participate in the improvement of their working conditions should be reinforced at both the national and enterprise level.

Source: Forastieri (2000), *Safe work: information note on women workers and gender issues on occupational safety and health*, International Labour Office, Geneva (http://www.ilo.org/public/english/protection/safework/gender/womenwk.htm).

ANNEX 9: SUMMARY OF THE RELATIONSHIP BETWEEN GENDER, OCCUPATIONAL STATUS AND WORKING CONDITIONS: FINDINGS OF A GENDER ANALYSIS OF THE THIRD EUROPEAN SURVEY ON WORKING CONDITIONS

Fagan and Burchell (2002) looked at the relationship between gender, occupational status and various working conditions indicators that are available from the three European working conditions surveys. The summary table they made of the results is given below. They conclude that analysis has shown that there are some gender differences in some aspects of working conditions, but no systematic pattern in all the indicators investigated.

Working conditions indicators	Gender difference	Relationship with occupational status group
Job content — selected items • Problem solving and learning • Task complexity/monotony • Team working/task rotation • Planning responsibilities	Full-time/part-time status has more of an impact than gender per se: part-time jobs offer fewer opportunities for learning, are more monotonous and have fewer planning responsibilities.	Problem solving, task complexity and planning are mainly features of managerial and professional jobs. In each occupational status group, these job content items are generally less prevalent for employed women compared to employed men. The smallest gender difference is found among clerical and service workers.
Skill match and the amount of training provided by employers • Skill levels match job demands? • Number of days training received?	A similarly large majority of each sex said that their skills match their job demands, and the amount of workplace training was also similar on average. Part-timers are the most likely to say that their skills are underused, and receive less training than full-timers.	Managers and professionals received more training than other workers. Within each occupational status group, women received less training than men; this 'gender gap' was least pronounced in the managerial and professional categories.
Customer service and 'people' work • The proportion of work time that is spent dealing directly with people external to the workplace (customers, passengers, patients, pupils, etc.).	Particularly a feature of women's jobs, whether in full-time or part-time work.	Particularly a feature of white-collar work. Women in managerial and professional jobs do more 'people' work than men employed in these occupational categories. The picture is reversed for blue-collar (manual) work: women have less 'people' work than men in these jobs.
Exposure to intimidation and discrimination at the workplace • Violence, intimidation and bullying • Discrimination (sex, age, nationality, ethnic background, disability, sexual orientation)	Women have higher rates of experience and awareness of violence, intimidation and discrimination at work.	
Exposure to material and physical hazards As a proportion of work time: • Loud noise • Vibrations from tools and machinery • Extreme temperatures • Dangerous products or substances	Men are more exposed to these hazards than women. Part-time jobs offer some protection against these hazards, both in terms of lower risks of exposure and shorter working hours.	Within each occupational status group, men are more exposed to these hazards. This gender difference is most pronounced in blue-collar (manual) jobs; it is negligible in the professional category. There is an interaction between gender and occupational status: the highest exposure is for men in blue-collar jobs, followed by women in blue-collar jobs; exposure is lowest for white-collar workers of either sex.
Exposure to ergonomic hazards As a proportion of work time: • Repetitive hand or arm movements • Painful or tiring positions • Carrying or moving heavy loads • Short repetitive tasks	No gender difference in rates of repetitive movement, painful positions, or short repetitive tasks. Men are more exposed to heavy loads, making them more at risk from these hazards overall. Part-time jobs offer some protection against these hazards.	Ergonomic risks are highest in blue-collar (manual) jobs. There is an interaction between gender and occupational status: women are more at risk of ergonomic hazards than men in all white-collar areas of work — particularly in professional jobs — as well as in blue-collar (manual) craft jobs.

Working conditions indicators	Gender difference	Relationship with occupational status group
Work with computers As a proportion of work time: • Work with computers • Telework from home with a computer	No gender difference in working with computers, but men are slightly more likely to telework. Part-timers are less likely to work with computers.	Gender differences are more pronounced when occupational status is taken into account: in each occupational status group, men are more likely to work with computers (including teleworking).
Homework • Home is the main place of work (excludes telework).	No gender difference in the rate of homeworking; no difference by full-time/part-time status (but the survey may underestimate women's greater involvement in casual or informal sector homeworking).	There are gender differences in the occupational profile of homeworkers. Male homeworkers are more likely to be managers or blue-collar craft workers while female homeworkers are more likely to be clerical or service workers.
Work intensity • Factors setting the pace of work • Working at high speed • Insufficient time to get the job done • Tight deadlines	Women are more likely to have their pace of work set by the demands of other people; men are more exposed to production targets or machine speed. No gender difference in the requirement to work at high speed or whether they have insufficient time to do the job. Men are more likely to work to tight deadlines. Part-time jobs offer some protection against work intensity.	Managers and professionals are most at risk from work intensity (but blue-collar workers are the most exposed to working at high speed). There is an interaction between gender and occupational status: among managerial and blue-collar craft jobs, women are less exposed to high levels of work intensity, but there are few gender differences in the other occupational status groups.
Disruptive interruptions	No difference by gender or full-time/part-time status.	The incidence is slightly higher in professional and managerial jobs.
Job autonomy and working-time autonomy • Work methods • Speed of work • Task order • When breaks are taken • Choice when to take holidays • Influence over working hours	Men have higher levels of autonomy. There are few differences between full-timers and part-timers in work autonomy, but part-timers have slightly more influence over their working time.	Autonomy is highest for managerial and professional workers. The work autonomy of women professionals is much lower than that of male professionals.
Wages • Net monthly earnings • Wage structure	Women are more at risk of low pay than men and are also less likely to benefit from the highest earnings. This gender difference is also found among part-timers. A higher proportion of men have additional payments built into their wage structure.	
Consultation about work organisation and workplace health and safety protective measures • Consultation occurs and is effective • Awareness of health and safety risks	The majority of men and women consider that they are effectively consulted, but one in three of each sex is not consulted. Among those working in hazardous conditions, women are more likely to report that they have insufficient health and safety information and they are less likely to have protective equipment.	

Source: Fagan and Burchell (2002), *Gender, jobs and working conditions in the European Union*, European Foundation for the Improvement of Living and Working Conditions.

ANNEX 10: MEMBER STATE DIFFERENCES IN SOCIAL SECURITY SICKNESS ABSENCE AND DISABILITY REGULATIONS

EU countries and social security regulations for sickness absence

Gründemann and Van Vuuren (1997) summarised the official security regulations for absence due to illness and disability in the Member States of the EU (and Norway). They concluded that great differences exist. However, it should also be noted that general practice often differs substantially from the official regulations.

In most countries (Austria, Belgium, Finland, France, Germany, Greece, Italy, Luxembourg, Norway, Portugal, Spain, Sweden and the UK), a certificate from a general practitioner is requested in cases of temporary sick leave. The situation varies somewhat in the different countries as to the number of days absence after which such a certificate is requested. In only three countries (Denmark, Ireland and the Netherlands) is such a certificate not requested. The requirement to produce a doctor's certificate in the case of sickness absence is usually intended as a threshold in order to make 'reporting sick' less easy. In practice, a medical certificate does not mean much. Employees will go to a doctor whom they know will sign a certificate, and, if that doctor will not cooperate, they can always go to another. Nevertheless, research shows that the duty to produce a medical certificate is generally linked to a somewhat lower incidence of sickness absence (see Gründemann and Van Vuuren, 1997).

In most countries, the worker who reports sick has to wait from one to a few days until he/she receives benefits. This is called the 'waiting period'. Only in Austria, Germany, Luxembourg and Norway are no waiting days applied. In the other countries, workers are not paid for the first day (Belgium and Sweden), the first two days (the Netherlands), the first three days (France, Greece, Ireland, Italy, Portugal, Spain and the UK), or the first nine days (Finland). In many countries, practice may differ a lot from these formal regulations. In many cases, employers compensate for the wage loss due to waiting days. Waiting days are, however, also used as a threshold in absenteeism. The literature, and the fact that practice sometimes is different from the regulations per se, shows that waiting days are accompanied by a lower frequency of absenteeism, but that the average length of the absence is greater. The effect of the differences in waiting days between the EU countries on the total absence will therefore be considered to be nil.

In most countries, there officially is a loss of income in the case of sickness absence, in the sense that salaries are not paid or that the benefit percentage paid is less than 100 %. Practice is usually less negative than the official rules would suppose. In many countries, the employer tops up the benefit, in many cases even up to 100 %.

After the first period of absence, the percentage benefit decreases in most countries. There is a continuation of full payment of wages in only four countries (Denmark, Luxembourg, the Netherlands and Norway). In the other countries, this percentage is between 50 and 80 % of the last-earned wage.

Most countries (11) operate with a maximum period of temporary unfitness for work of approximately one year. In Italy and the UK, a shorter maximum period is applied (26 and 28 weeks, respectively). The maximum period of absence has an important influence on the length of the sick leave, particularly on the registered sick leave. It may, however, also influence the attitude towards rehabilitation, and thus also influence the total days of absence.

Although all EU countries have regulations for maternity leave, the duration varies greatly from country to country. Particularly in sickness absence registration, maternity leave may be responsible for a significant part of the sickness absence. For example, in the Netherlands healthcare system, an average of 1 to 1.5 % of work time is taken up by maternity leave (Gründemann and Van Vuuren, 1997).

Long-term or permanent disability regulations in the EU

In many countries (Belgium, Finland, Ireland, Luxembourg, the Netherlands, Norway, Portugal and Spain), the regulations on extended or permanent disability are linked, timewise, to regulations governing temporary sick leave. In these countries, there is a waiting period for the permanent disability benefits, which is equivalent to the maximum period applicable to (non-permanent) sickness absence. In most other countries, it is not essential for the maximum period for (non-permanent) sick leave to have elapsed before a person is entitled to permanent benefit (Gründemann and Van Vuuren, 1997).

The definitions and conditions of payment are diverse. They are often based on a minimum loss of earning capacity or a minimum percentage of unfitness for work. This minimum may differ considerably between countries, as may the maximum benefit. In Belgium, the benefit percentage is dependent on family circumstances. In most countries, there is an additional assessment procedure, which is decisive in the assignment of disability benefits or invalidity pension. Here, differences are also found between countries (Gründemann and Van Vuuren, 1997).

ANNEX 11: RESULTS FROM THE TUTB SURVEY ON THE GENDER DIMENSION IN HEALTH AND SAFETY, AND EXAMPLES OF TRADE UNION INITIATIVES

The TUTB survey on gender and OSH

In order to facilitate the incorporation of gender into their workplace health and safety policy, the Trade Union Technical Bureau (TUTB) and European Trade Union Confederation (ETUC) carried out a survey in the 15 EU countries in 2001–02. The survey, supported by the Belgian Ministry of Employment, in connection with activities pursued by Belgium's EU Presidency, was to assess:

- the inclusion of gender issues in health and safety policies;
- health and safety interventions at the workplace that take account of gender issues.

Respondents included: trade unions (36 %); research institutes (21 %); institutions responsible for occupational safety and health policies (13 %); and preventive services (9 %). Very few replies were received from institutions responsible for equalities policies. The survey was complemented by a search and analyses of how articles in scientific or informational occupational safety and health magazines dealt with gender and a broader literature search.

The initial findings include the following:
- The gender dimension is gaining recognition as a factor in workplace health and safety.
- Issues covered range from musculoskeletal disorders to the organisation of working time, and across traditionally male strongholds like the construction industry to female-dominated occupations like nursing and cleaning services.
- Some sectors are clearly much further on than others in this area: 25 % of activities were in the health and social services sector; 10 % were in distribution and retail (e.g. supermarket checkouts); but less than 25 % were in industry, mainly in the textile, footwear and clothing sectors.
- Most occupational safety and health policies are still based on a gender-neutral model, historically based on the standard male worker and generally occupational safety and health institutions ignore gender, with the exception of the Nordic countries.
- Prevention services are beginning to integrate the gender dimension into their activities, but generally only in 'women-dominated' employment sectors, or with regard to problems considered to be 'more specifically for women'.
- Research suffers from policy compartmentalisation; for example, there is significant research on gender segregation in employment, but little of it addresses occupational safety and health issues; many 'time budget' surveys have looked at division of time between activities but few have linked this to working conditions.
- There are different interpretations of what constitutes gender-sensitive research. It may be interpreted as research: focusing on female-dominated jobs; including a com-

parative analysis of women and men; focusing on issues considered to be particularly relevant to women, for example reproductive health or work–life balance. Other research considers the gender dimension in factors both inside and outside the workplace, including the study of male populations from a gender perspective.
- Different fields of research have different approaches. None guarantees coverage of the gender dimension. An interdisciplinary approach is needed, that also covers life time and work time ('cross-cutting approach'). Taking account of the subjective experience of workers is very important.
- Only a minority of occupational safety and health research covering gender is aimed at interventions and prevention.
- Health at work, equality and social health policies are very compartmentalised and relatively impervious to issues in the others.
- Health at work policies tend to disregard the interaction between paid and unpaid work. If gender is covered, it is often as women only, who are labelled as a 'vulnerable group'.
- Public health policies have become more gender aware in recent years. However, neither paid nor unpaid work features highly in most studies of the gender dimension of health. Where the link between health and unpaid work has been considered, the focus has been more on access to employment than to explore the link of the dual workload on health.
- Public health usually brings working conditions into the equation only when there is an immediate link between a particular factor and a medical condition. This affects women more, for example because women's work-related health outcomes are more likely to be multi-causal than men's.
- Health policies at work are not joined with equality policies and there is a low visibility of occupational safety and health in the employment equalities field.
- In many countries, positive policies are pursed to promote gender balance at work; however, most reported cases do not cover changing working conditions, but are limited to vocational training, sometimes linked to psychological support.

Differences by country were also observed, particularly between Scandinavian and Latin countries. In the Scandinavian countries, the TUTB found the gender dimension to be more visible and integrated into institutional policy and that there appeared to be a greater number of OSH and gender studies taking place. Gender indicators were used in monitoring activities, and specific programmes and cooperation between various players had resulted in a more systematic stock of knowledge. The gender approach had been applied to a broader range of health problems in both research and statistical monitoring. In the Latin countries, the TUTB found that there was less institutionalised activity on gender and OSH, although it had increased in recent years, and that, where it took place, it was more likely to be limited to 'women-specific' problems. However, as there was still a high 'social' demand for activity on gender and OSH, this was resulting in action taking place at a local level. The UK was described as having a pragmatic approach which does not explicitly cover all gender and OSH issues, but which has given rise to both local initiatives, generally instigated by trade unions, and research programmes by the OSH authority that have become more gender-sensitive.

Sources: Vogel (2002); TUTB web site.

Examples of trade union actions on gender and OSH

The TUTB has also collected examples of trade union initiatives aimed at prevention of hazards to women and improving women's participation. For example, Milan's three trade union confederations, CGIL, CISL and UIL, set up a women's occupational health task force of trade unionists, public prevention service technicians, doctors, and workers' safety representatives in 1996. Its areas of study have included repetitive work by women in different industry and service sectors, as well as biological risks, maternity protection and night work.

The Spanish Trade Union, CCOO, obtained an agreement with the Balearic Islands' biggest hotel chain to set up an information and trade union training initiative on musculoskeletal disorders in the company's 30-odd hotels. They developed a training programme for prevention representatives, aimed at women hotel room-cleaning staff, whose main problems are backache and other musculoskeletal disorders. The training, carried out by the union, takes place on the hotel premises. It covers recognition of musculoskeletal disorders, risk assessment and drawing up a prevention plan. A survey of room-cleaning staff is also made. Proposals made by the workers throughout the course and from the cleaning staff survey are collected and incorporated into the prevention plans. The safety committee takes forward the resultant plans and they form the basis of discussions with prevention service technicians and for preventive action. One important outcome was that room cleaners had previously played little part in safety committees or in purchasing decisions (one issue was poor ergonomic design of furniture and work equipment).

Source: TUTB (2000).

ANNEX 12: GENDER EQUALITY LEGISLATION AND POLICY IN THE EU

The EC Treaty puts the promotion of gender equality among the tasks of the Community (Article 2) and makes it a transversal objective (Article 3) for the EU. Article 13 of the EC Treaty entitles the Commission to take initiatives to combat discrimination based, among other grounds, on sex. Article 141 is the legal basis for Community measures for equal opportunities and equal treatment of men and women in matters of employment. The Commission's gender equality web site (http://europa.eu.int/comm/employment_social/equ_opp/rights_en.html#other) has compiled links on legislation and other actions on gender equality, and links to the directives and acts listed below can be found there.

EU directives on gender equality cover the following themes:
- Equal pay (Council Directive 75/117/EEC)
- Equal treatment at the workplace (Council Directive 76/207/EEC)
- Equal treatment with regard to statutory social security schemes (Council Directive 79/7/EEC)
- Equal treatment with regard to occupational social security schemes (Council Directive 86/378/EEC)
- Equal treatment for the self-employed and their assisting spouses (Council Directive 86/613/EEC)
- Maternity leave (Council Directive 92/85/EEC)
- Organisation of working time (Council Directive 93/104/EC)
- Parental leave (Council Directive 96/34/EC)
- Burden of proof in sex discrimination cases (Council Directive 97/80/EC)
- Framework agreement on part-time work (Council Directive 97/81/EC)

Other Community acts include coverage of the following:
- Gender balance in decision-making
- Reconciliation of work and family life
- Development cooperation
- Women and science
- Violence and sexual exploitation
- Education and training
- Equal pay
- Employment and labour market
- Structural Funds
- Human rights and multiple discrimination

ANNEX 13: CRITERIA FOR ASSESSING WORKPLACE EQUALITY

A self-assessment checklist for workplace equality is given below. Equality is assessed against eight separate standards.

The level of equality	✓ Are equality and staff well-being incorporated into human resource management? ✓ Is equality inherent in the goals and strategy of the work organisation? ✓ How is equality perceived by the staff members?
Salary and remuneration policy	✓ Is the salary and remuneration policy based on equal treatment and fairness? ✓ How openly are the bases for remuneration and bonuses presented and discussed? ✓ Do people feel that they are treated justly?
Career and work opportunities	✓ Do people have equal job opportunities and equal opportunities for career advancement and lifelong development in their work?
Common goals and opportunities for influence and control	✓ Is everyone familiar with the common vision and common goals?
Work atmosphere and feeling of togetherness	✓ Does the workplace culture support equality? ✓ Is diversity seen as richness, or is the goal a homogeneous staff?
Information flow and openness in information delivery	✓ Does everyone have equal opportunities regarding information concerning his/her own work and work unit and future conditions and perspectives, also in economic terms?
Working conditions	✓ Are the working conditions safe and ergonomically good? ✓ Is age management practised and older people given due attention? ✓ Have work ability programmes been initiated, and do they support staff well-being?
Reconciling work and family life (private life)	✓ Does everybody have an opportunity for private life outside work? ✓ Is continuous overwork and overcommitment required and rewarded? ✓ Is refusing overwork being punished? ✓ Are a personal life and family life valued, or is family seen as a burden? ✓ Are men being supported and rewarded, or discouraged and subtly punished for taking parental leave?

Source: Kauppinen and Otala (1999).

ANNEX 14: QUALITY OF WOMEN'S WORK: STRATEGIES FOR CHANGE IDENTIFIED BY THE EUROPEAN FOUNDATION

It was recognised at the beginning of this report that a broad range of issues relating to equality in employment as well as in social and civil life have an impact on gender differences in occupational safety and health in the workplace itself. Investigating these issues is largely outside the scope of this report, and others at the EU level have put forward a strategy for action in this area. The following proposals and examples of good practices have been put forward by the European Foundation (2002c), based on its research.

1. Areas for collective bargaining

- Pay discrimination/pay equity:
  Pay levels, opportunities, systems, pay and grading structures, evaluation of jobs, access to benefits, etc.
- Gender segregation:
  Access to/nature of training, recruitment, promotion, job definitions and qualifications, work organisation, restrictions on women working, etc.
- Job access/job security:
  Redundancy, termination, security of hours, contractual status, etc.
- Family–work interface:
  Maternity/paternity/parental and dependent leave, childcare, working time, etc.
- Organisational cultures/structures:
  Equal opportunities/awareness training, career paths, sexual harassment, etc.

(Bercusson and Dickens, 1996)

2. Tackling gender gaps: examples of innovative agreements

The following are examples of innovative features of agreements directed at improving women's job access, reducing sex segregation and diminishing pay discrimination to contribute to closing gender gaps in employment.

Improving women's job access and career progress

- Elimination of sex stereotyping in job descriptions and advertisements.
- Opportunities to combine work and caring and flexible time arrangements.
- Removal or raising of age limits and the elimination of discriminatory requests for information.
- Positive action recruitment advertising (to encourage applications from the under-represented sex).
- Setting of recruitment targets.
- Checking for suitable internal candidates and inviting all female candidates for interview or putting female candidates on the shortlist for jobs in which they are under-represented, at least in proportion to the number of women among the applicants.

Promotion and training

- Commissioning studies of the gender composition of the workforce.

- Identifying obstacles to the promotion of women.
- Mapping of career paths to facilitate access by women to higher posts.
- Equal or preferential access to training and work experience.
- Special training (for example, enabling women to acquire 'male' skills and for managers and others in equal opportunities awareness).
- Training funds and places reserved for women.
- Arrangements for care facilities during training.

Closing the gender pay gap

- Reviewing the context in which the agreement is to be implemented (for example, making sure that all employees are covered by the agreement — whether working part-time or on a non-permanent basis).
- Developing new tools for gender-neutral job evaluation.

Reconciling work and family life

- Training measures during parental leave and in relation to reintegration into employment.
- Maintaining contact while on leave.
- Building up seniority rights and social security rights during periods of leave.

3. Contextual factors influencing the outcome of actions and interventions

The Foundation's research also highlighted the impact the overall context has on collective bargaining for equality (Dickens, 1998). The following factors were identified.

- **Environmental factors:** economic context, labour market, legislative and non-legislative interventions.
- **Organisational factors:** employers' interests and concerns, unions' interests and concerns, facilitating internal contexts, human resource policy.
- **The significance of gender in collective bargaining:** the identity of the negotiators, the importance of women's presence, nature and quality of the bargaining relationship.

4. Issues identified for future research and initiatives

Within the context of an enlarging European Union, the Foundation's research has identified a number of issues for future research initiatives:

- More in-depth country comparative research.
- Developing gender-sensitive indicators on quality of work and employment (status, occupation, income levels and wages, social protection, health and safety, work organisation, competence development, work–life balance).
- In-depth analysis of the factors behind gender segregation of labour markets in the EU and of the strategies in place to address this issue.
- Further research on how to balance working and non-working time with a life-course perspective.
- Relationships between paid and unpaid work and their links to social protection and pensions.
- The quality of women's work and employment in the candidate countries where there seems to be higher participation rates of women and less gender segregation than in the EU.

Source: European Foundation (2002c), *Quality of women's work and employment — Tools for change.*

European Agency for Safety and Health at Work

Gender issues in safety and health at work — A review

Luxembourg: Office for Official Publications of the European Communities

2003 — 222 pp. — 16.2 x 22.9 cm

ISBN 92-9191-045-7

Price (excluding VAT) in Luxembourg: EUR 25